CURING CORAL

A Marine Biologist's Guide
to Coral Reefs and
How They Help Us
Combat Climate Change

Summer Collins

mango
PUBLISHING

MIAMI

Cover, Layout & Design: Megan Werner
Cover Illustration: Yelyzaveta / stock.adobe.com
Interior Illustrations: venimo / Екатерина Заносиенко

For permission requests, please contact the publisher at:
Mango Publishing Group
5966 South Dixie Highway, Suite 300
Miami, FL 33143
info@mango.bz

For special orders, quantity sales, course adoptions and corporate
sales, please email the publisher at sales@mango.bz. For trade
and wholesale sales, please contact Ingram Publisher Services at
customer.service@ingramcontent.com or +1.800.509.4887.

Curing Coral: A Marine Biologist's Guide to Coral Reefs and How
They Help Us Combat Climate Change

Library of Congress Cataloging-in-Publication number:
2025938408
ISBN: (pb) 978-1-68481-818-1 (e) 978-1-68481-806-8
BISAC category code: NAT011000 NATURE / Environmental
Conservation & Protection

Some names have been changed to protect the privacy of
those depicted.

CURING

CORAL

Praise for
Curing Coral

"This is an exceptional read for all audiences, filled with relatability, valuable insights, and captivating storytelling into the impact climate change is having on our coral reefs and planet."

—Veronica Williamson, marine biologist

"A must-read! I couldn't be prouder of the awareness that Summer is bringing to coral reef conservation."

—Zachary Craig, Hawai'i coral restoration coordinator

"An important look into the good, the bad, and the ugly of being a marine biologist fighting our climate crisis. Summer exemplifies the resilience and perseverance necessary to make it in an ocean world full of joy and devastation. An essential read for any aspiring marine scientist."

—Megan Kennedy, senior research associate in the Coral Reef Restoration Lab at the University of Miami and program manager at Rescue a Reef

"Both educational and entertaining, *Curing Coral* is an excellent read for anyone interested in ways to save our oceans."

—Bonnie Monteleone, co-founder of Plastic Ocean Project Inc.

"There is so much knowledge and passion for the ocean within these pages and you can hear it through the writing. I felt like I was right alongside Summer on all her adventures, laughing and aw-ing with every other sentence I read. This book made me have so much more love and admiration for the coral reefs!"

—Chloe Spring, marine biologist

To the people I love who have helped me along the way. You know who you are.

To anyone who works for the planet, I see you. Thank you for what you do.

"Every time I slip into the ocean, it's like going home."

—Sylvia Earle

"Even the tiniest creature on Earth has a purpose. There is no such thing as an insignificant life."

—Eugenie Clark

Table of Contents

Introduction | What Is It like Being a Marine Biologist?..............13

Chapter 1 | Coral Reefs: The Breakdown..............17

Chapter 2 | Belize..............21

Chapter 3 | Guidance..............29

Chapter 4 | Turtle Talk..............37

Chapter 5 | Just Say Yes..............55

Chapter 6 | Attack on All Sides..............69

Chapter 7 | Multifaceted Problems Equal Technically
Advanced Answers..............85

Chapter 8 | The Coral Restoration Program..............95

Chapter 9 | Intern Turned Mentor..............109

Chapter 10 | Living Rocks and Their Population Problem..............119

Chapter 11 | Baptism by Fire..............135

Chapter 12 | Coral World..............147

Chapter 13 | Hell Week..............155

Chapter 14 | Here's to Hope..............167

Chapter 15 | Caring About Coral..............173

P.S. 179

References 181

Acknowledgments 191

About the Author 195

Introduction

What Is It like Being a Marine Biologist?

"So, what do you do for work?" The inevitable question gets asked every time I meet someone. Whatever possible social situation you can think of—attending weddings, birthday parties, flights, holiday get-togethers—I commonly find myself being on the receiving end of this inquiry. Most people have pretty vague answers, eager to avoid thinking about their job while not on the clock. My husband, whom I've been with since high school, always swells up with pride at my answer. "Marine biologist," I respond with a grin and a shrug. This is usually followed by a flurry of additional questions and enthusiastic interest that has me talking much longer than I ever anticipated. Almost everyone I know has thought of being a marine biologist at one point or another. A normal response after I mention what I do for work is "I've always wanted to be a marine biologist!" The everlasting horizon and deep blue create an allure that resonates with many people. This allure either lands them in the category of awestruck by or frightened of our world's oceans, but curious nonetheless.

My friends and I find ourselves in the former category (thankfully—otherwise that'd be weird). Granted, I am friends with a bunch of nature nerds, but that's beside the point. While being a marine biologist truly is the cat's pajamas, not everyone knows what it's actually like being a marine biologist. Therefore, after a lifelong dream, a reading obsession, and prolonged interest from my peers, I've decided to write about it. Here's the story of how a Florida girl had a dream of doing shark research and ended up in the fight to cure coral instead. I'm writing this to share my story. To all the future marine scientists out there. To inspire change in the hopes of a better future for our oceans and our planet. To the dreamers of a better world.

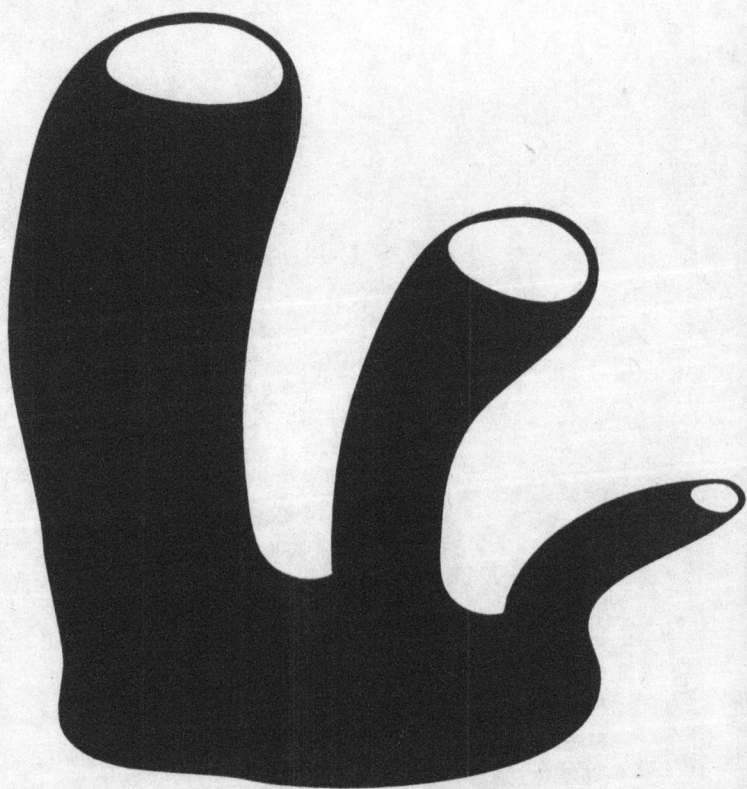

Chapter 1

Coral Reefs:
The Breakdown

S o, what is a coral reef? I'm sure you're thinking of the vivid images flashing across your TV screen, showing ultra-colorful landscapes and fish darting about. You would not be wrong. However, a coral reef is more than just a pretty video, a secret fishing hole, or a good spot to snorkel on a vacation. A coral reef is an ancient civilization made up of hundreds of thousands of different animals. It is an entire ecosystem, comparable to our rainforests. These animals create a labyrinth of structures and color, playing a symphony of crackles and pops, providing an oasis on the ocean floor. These foundational animals called coral lay out the base which every animal builds upon. Fish, dolphins, sharks, crabs, and many more creatures rely on coral reefs for food and habitats. Outside of the animal realm, coral reefs have been the inspiration for writers, artists, and scientists throughout time.

Coral reefs form in warm shallow waters where sunlight can still penetrate the surface. On average, they exist anywhere between sixty to ninety meters down, but can reach as deep as 150 meters. The warm coastal waters of places like Australia, Florida, Costa Rica, and the Maldives offer the perfect amount of sunlight and water temperatures of 73–84 degrees Fahrenheit for these ancient cities to form. Coral reefs began forming millions of years ago, with the oldest carbon dating of corals pointing to over 500 million years ago. There were also recorded notes of corals, particularly red corals, in ancient civilizations like Greece and Rome. Even Aristotle mentioned them in his biological studies!

However, our understanding of these incredible ecosystems only really began to develop between the seventeenth and eighteenth centuries, most noticeably by Charles Darwin's written observations in the Galapagos when researchers first began to recognize corals as animals rather than plants. Henri Milne-Edwards and Jules Haime conducted significant work toward this taxonomic classification by studying their polyps and skeletal structure, though that was even further into the nineteenth century. Before we get too far into individual corals themselves, you'll need to understand the formation of the reef as a whole. There are four different kinds of coral reefs: fringing reefs, atolls, barrier reefs, and patch reefs, each with their own distinct patterns and characteristics.

Fringing reefs are the most common type of reef and are directly attached to a shore or bordering a coastline. They grow outward from the shore and are separated from it by a shallow lagoon or reef flat. The proximity to land can expose fringing reefs to more sediment and freshwater runoff, which can affect their health.

Barrier reefs are similar to fringing reefs, but are separated from the shore by a deeper and wider lagoon. The most famous example of a barrier reef is the Great Barrier Reef in Australia. Barrier reefs form parallel to the coastline but can be several kilometers offshore, creating a substantial barrier between the open sea and the shore.

Atolls are ring-shaped reefs that encircle a lagoon, often without any central island. They typically form around the subsiding remnants of volcanic islands. As the island sinks, the coral continues to grow upward, maintaining the reef structure above sea level. Atolls are common in the Indo-Pacific region.

Patch reefs are small, isolated reefs that grow up from the open bottom of the continental shelf or lagoon, often between larger reef structures. They can vary greatly in size and are usually found within lagoon systems associated with barrier reefs or atolls.

My work currently takes place in the Florida Keys. However, before I ended up here, the first functioning reef I visited was in Belize.

Chapter 2

Belize

I leaned over the boat to puke. In time, I would come to think of it as calling for dinosaurs, given the atrocious sounds wrenched from me as I unwillingly gave up my lunch. In the moment, all I felt were waves of embarrassment from being ill while surrounded by my travel abroad group. We could not make it there fast enough; two hours was a demanding boat ride. In an attempt to ease my nausea, I looked out to the horizon. Sweet relief washed over me in my state of unbalanced equilibrium as the island came into sight. The island, nothing more than a speck, was one mile of sand off the coast of Belize and home to a research station, two stray dogs, a crocodile, and the coastguard crew. This would be my home for the next ten days—and an introduction to a passion that would last a lifetime.

In the marine science world, what draws us to Belize is the reefs. One of the most famous spots you may know of in this area is the Blue Hole. Beyond that, Belize is actually home to the second-largest reef tract on the entire globe, making it the perfect location for a group of twenty college students and two brave professors there to teach them the ropes of field work and research. Little did I know that I would become a coral biologist!

The next morning, our professors had us break into groups on the research vessel and extract our snorkel gear from our bags. The reflective aqua surface of the water beckoning me, I plunged into the most ancient of all civilizations. Millimeter upon millimeter, year after year, millions of coral polyps had laid the groundwork to build the intricate interlocking structures providing refuge for a quarter of all marine life. A serenade of crackles greeted my ears as I dove deeper under the surface, eager to explore this bustling city. A patchwork of colors, structures, and textures wove together to create one of the most unique environments on the planet.

Among the large boulder corals, I noticed the largest porcupine fish I'd ever seen gazing up at me, galaxies swirling in its eyes. I turned as a breeze of yellow snapper cruised past on the gentle current. Fighting the burn in my lungs, my eyes drank in more of the sights. As I rose to the surface, the elkhorn colony captured my attention, standing proud amidst the underwater metropolis. This was infinitely better than any aquarium, any at-home saltwater tank, and any other Florida reef that I had ever seen. Every marine biologist says this, but I wanted to grow gills. I didn't want to come back to the surface. *Just leave me with the fishes*, I thought. Day after day, we would visit the reef in these groups. Every outing was accompanied by classroom lessons on species identification, how to take a transect (a fixed path that goes through a natural landscape in order to create standardized observations), underwater navigation, and so on. Every time I went back to the underwater city, I was greeted with new discoveries, childlike wonder, and marine life magic. Until I wasn't.

The purpose of our trip was to collect data to add to the ever-growing sample that was cataloged with each year's travel abroad cohort. We were there to collect data on things like sea grass species and length, mangrove species and diameter, and to analyze the local reefs. The intricacy of the local reefs made for a broad number of topics to be investigated. Trips handled spawning, species dispersion, fish to coral interactions...the list seemed endless. My research focus was made abundantly clear when I came across a graveyard.

At the edge of the reef, it was quiet. The surrounding area was in mourning for the loss. The tones of yellows, browns, and reds had been replaced by a void of color. White. Everything was white. Ancient pillars crumbled into ruin on the ocean floor. Pieces of history, chunks of their skeletons rolling on the bottom with the ocean's surge like chalk. It was so out of place. I looked out at the rest of the reef, shaking my head. This wasn't supposed to be here. Time to break out the transects. My study team and I proceeded to take ten sets of videos, each five meters long, in three different reef locations. We then analyzed the video footage and counted how many coral colonies we saw that were bleached or partially bleached.

Corals have this tiny microorganism in their tissues called zooxanthellae. Zooxanthellae (pronounced zoh-uh-zan-thel-ee) is a type of algae that lives within the coral tissues. Think about it like tenants living in an apartment building. These tenants are very good to their host, as they provide food and energy to the coral for letting them live there. We call this relationship between zooxanthellae and corals a symbiotic one. Zooxanthellae get a lovely apartment building, and corals get regular meals and energy supplied to them. Whenever corals get too hot, they kick out their food-making tenants. Corals ideally thrive in temperatures between 73–84 degrees Fahrenheit (23–29 degrees Celsius). They can tolerate up to 32 degrees Celsius. Recently, in June 2023, our ocean temperatures rose to 101 degrees F (38 degrees C). Legally, if a landlord's

AC cuts off, they have to evict everyone from the building; the same thing happens between corals and zooxanthellae. This sudden rise in sea temperature is the same as cutting off that AC, and the zooxanthellae are booted out. Luckily, corals can survive and recover from this event and invite the zooxanthellae back in, but only if the AC gets turned back on. In other words, only if ocean temperatures drop back to normal. Without their zooxanthellae counterparts, coral slowly starve and will eventually die if they cannot get their tenants back. A horrible way to go, really. No one wants to starve to death.

We chose three different reefs to work on: Half-moon key, Blackbird key, and Calabash reef. We snorkeled every reef, measuring out a sixty-foot transect for a video recorder to take a video swimming along the transect. We ensured we took all the videos at the same depth and the same distance at each location. We proceeded to take five videos at each reef. Once we were back in the States, we analyzed the footage using a simplistic analysis. Each member of the five-person team paused the video at various points, counting each coral in the frame that displayed a form of coral bleaching. We separated these recordings into partially bleached, half-bleached, and entire colonies bleached. We wrote down our observations into an Excel spreadsheet that we then used for fancy statistical analysis to calculate the percentage of bleaching recorded at each reef location. We repeated each step of this process for all of the reefs we visited. Once we were back on US soil, covered in about seventy bug bites each, we returned to our statistical analysis and drafted up our presentation to the class about our findings. Each group presented on a variety of topics: seagrass growth analysis over the past ten years, mangrove species dispersion on the island, and more. We were the only group to attempt to analyze coral bleaching.

At the time, it didn't hit me just how impactful that trip was. I had always envisioned coral to be interesting, but I wasn't sold on that being the focus of my career. In the meantime, while I was busy focusing on graduating, I explored other things. I always had an odd sense, though, seeing that many of my peers knew exactly what they were going to do, their entire careers and steps planned out year by year, and I didn't. This "odd man out" feeling left me constantly seeking advice and trying new things.

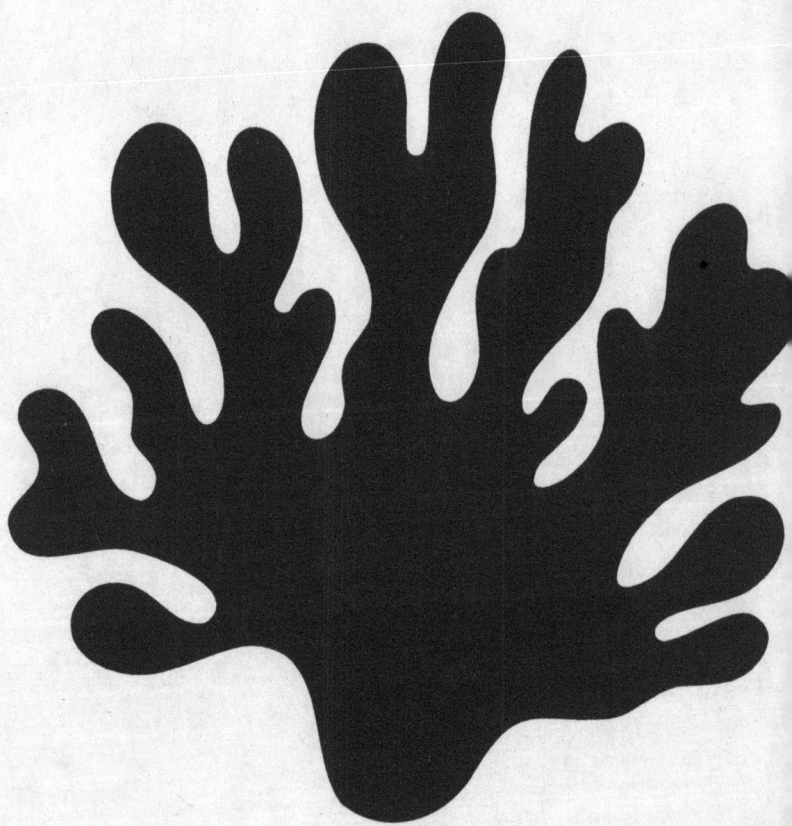

Chapter 3

Guidance

I hit snooze when the first alarm blared the next morning after the presentation. I blindly reached for the alarm, knocking my phone to the floor. "You've got to be kidding me," I grumbled to myself as I forced my legs to swing over the side of my bed. Rolling my shoulders, I coaxed myself to search for the blaring rectangle hidden under tossed bedsheets and clothes deemed unwearable. The phone read 7:25 a.m. and I nearly did a double take.

Spewing a stream of curse words, I panic-ran through my apartment, throwing on a pair of leggings and scrambling for a hair tie. I threw my curls into a bun and brushed my teeth as quickly as humanly possible. I had found a single bedroom apartment off campus and, with the help of my parents, moved into it earlier that year. Drawbacks? The apartment was fifteen minutes from campus. At 7:35 a.m. I was bolting down the stairs. I jumped into my black Volkswagen Beetle, nearly spilling the coffee in my cup holder, and floored it to campus.

All scientific labs were in a separate lab building close to the library, and I prayed there was a single parking spot so I could make it. By the grace of the education gods, there was a spot left. The door blissfully swung open to the lab room, where I was greeted with a bunch of students turning to me and interrupting the TA (teaching assistant) midsentence. Her eyes thinned at me in annoyance. "You may take your seat." Wobbly with relief and out of breath from running, I locked eyes with my lab friend, Jackie, who just smiled and gave me a thumbs-up in a "Congrats, you actually made it!" gesture. The specimen jars and microscopes on the table finally caught my attention and I tried to steady my breath.

"How do you think you did?" Jackie eyed me as we walked side by side leaving the lab exam. She had waited for me; I was one of the last remaining students.

"I like to always say that I failed, so when I don't, it's a pleasant surprise," I responded sarcastically.

She rolled her eyes, but grinned. "I'm sure you didn't fail." I nodded in agreement.

"I cannot believe she made us isolate and define each fin movement as well as their purpose on the *Balistes capriscus*!" I said as an afterthought.

"I know, right?!" she exclaimed.

We chitchatted back and forth as we made our way to our next destinations. The lab was the only class of my day, so instead of going home like a normal person, I decided to get some advice to address my conflicted feelings about the entirety of my career and what to do with my life. Dr. Jack Hall was the department head of the environmental science program at my university. He held the answers to most of my questions, even the philosophical ones. He was also one of the two brave professors to escort my group to and from Belize, and after attending his Introduction to Environmental Science course and the Belize course, his office was a recurring place I found myself in.

I'm going to break the fourth wall and talk directly to you, reader. Hi, hey, hello, salutations. I hope this book finds you well. If you're in college, about to be in college, or even working professionally, there's some pretty solid advice coming up, so you may want to grab a highlighter or a pen. Pick at least one person—this person should have more experience than you in whatever you're interested in—and learn from them. This person can be a professor, a boss, or a senior teammate. Then, make the effort to truly take advantage of that knowledge—you would be shocked by how much you miss out on simply because you're afraid to ask. All you have to do is ask in a genuine "You know more than me about [insert any topic you like] and I'm just a person who likes this stuff, please teach me the methods to your madness, oh Master Jedi" kind of way. You would be shocked where it could lead you and how it could open doors for you, land you new jobs, get you promotions—the whole gambit. And "cut!" on the fourth-wall breaking.

Jack, as I had come to know him, was in his early sixties but acted like he was still in his early twenties. Every day, he wore Hawaiian button-up shirts that were as bright and colorful as his boisterous personality. To top off his style choices, he never wore shoes, either. He told bad jokes on Thursdays in class and held a five-star rating on the "Rate My Professor" website, making him a local legend and prized professor on campus.

I entered his office, and his tall, lanky form was hunched over his desk, face pinched in concentration on a paper. I knocked on the frame of his door. "Bad timing, Dr. Jack?" I asked.

"Thank goodness. Here I was, about to start begging for a distraction. What's up, kid?" He also had a knack for calling every student "kid" or "kiddo," me being one of them.

I entered his office, which was about the size of a spare bedroom and full of books. He had stacks of them on either corner of his desk. Posters of sea life and a movie poster of *Jaws* decorated the walls.

"Just looking for general advice. I figured you might be crazy enough to give me some, but only if you have the time, of course," I said questioningly, refraining from plopping myself into the cushioned chair across from his desk.

Jack peered at me over his round spectacles, which slightly enlarged his eyes. "Are you going to stand the whole time or are you going to sit?"

I sat with a chuckle. "I take it that means you have time."

"Like I said, please save me from that horrendously-written paper I was grading," he retorted, so I began asking all of the questions that had been bubbling in my brain. I asked him what his story was and how he ended up being a professor advocating for saving the planet after working as a geologist locating oil fields. I asked him what it was like to scuba dive and how long it took to get different certifications. I asked him about work-life balance and why everyone in this field jokes that "work is life." I told him that I had made a simple decision about my career, but I wanted advice to see if it was actually a real thing.

He responded in kind, filling me in with stories about his life and his career for well over an hour. Pointing to the large stack of dive logs on his bookshelf, he told me that he had completed well over 2,000 logged dives and even worked in an underwater laboratory for a period of time. He shared how he almost died on several occasions, and how he, too, fell in love with the ocean.

That's when I asked the vital question: "Then why are you an environmental science professor?"

He leaned back farther in his chair with a knowing smile on his face. "The same reason you're probably going to switch from a marine biology degree to an environmental science degree. I wanted to work outside, and I didn't want to be holed up in a lab all day freezing my ass off."

And there it was. The small, unconscious decision that I had made, laid out before me.

We then proceeded to look at my courses and came up with a game plan to shift degrees. Miraculously, all of the courses that I had taken for marine biology also counted toward an environmental science degree, and all I needed was to switch out a few during my senior year.

The plan was in place, I knew what courses I needed to take, and everything seemed to be set.

"One last question. If I have an environmental science degree, how do I get a job in marine science?"

Jack responded, "Easy. All your hands-on experiences just need to be marine science-related. So, pick internships in accordance with jobs you might want to have. That's what an internship is, anyway: test running a job."

I hadn't heard internships described in such a straightforward way before. All I had heard was that you needed them to graduate, diversify your resume, demonstrate hands-on experience...blah blah blah. Jack's straightforward no-nonsense explanation had made it all glaringly obvious. So, that's exactly what I did. I went and test ran jobs.

Chapter 4

Turtle Talk

My palms had never been sweatier in my life. I rubbed them against my cotton skirt, hoping to remove the slick sweat found there. "You're going to do fantastic, you sound so professional," my best friend said, ushering me out of my dorm room with a smile and these words of encouragement. Kimmy, my classmate and best friend, actually showed me this internship at the local sea turtle rescue and rehabilitation center and said we should apply together.

I walked through the hallways of the biology building, taking deep breaths. "Professional...calm and professional," I muttered to myself. I wiped my hands against my skirt again.

I found the door to Dr. Keely's office. A paper taped to the dark wood read "Office hours by appointment only" in bright red, underlined, and italicized text, with several exclamation marks. Lovely.

I checked my phone: 3:55 p.m. Our meeting wasn't until four, but my father's words rang in my head: "*Early is on time, on time is late, and late is unacceptable.*" I stood in front of her door fighting the inner turmoil of what to do. *Do I knock now? Or at four o'clock sharp?* I wondered. I started biting my nails. *Screw it,* I decided, *she can be mad if she wants, it's better than looking stupid waiting outside of her door like this.* I reached up and gently knocked.

A moment later, a short woman with a brown bob peered up at me. "Yes, how may I help you?"

I introduced myself. "Hello Dr. Keely, my name is Summer Brooks, and I'm here for our four o'clock interview regarding the sea turtle internship?" The last part sounded more like a question than a statement. So much for confidence, I guess.

Dr. Keely's face brightened with a broad smile, and she proceeded to welcome me into her office. "Of course, yes! Please come in," she said, sounding much nicer than her office appointment warning.

Her office was a cozy space with a large window and a dark mahogany corner desk. She had an entire turtle skull sitting on the top, along with a variety of sea turtle posters scattered along her walls. My eyes swept to the couch and a coffee table on the lefthand side of the room.

"Please, sit!" she assured me. I rested my notebook on my lap, trying to look as professional as I could and not give away that I was actively crapping my pants.

She proceeded to explain a little more about the internship. "You'll be expected to be at the facility by seven every Saturday morning, and you'll be working anywhere from eight- to ten-hour shifts at the hospital. You'll be working with a variety of sea turtle species there, but they have some long-term rehabilitations that have been at the facility for a few years. There will be additional opportunities to sit on nests throughout the night if that's something you're interested in."

I perked up when I heard about the nest-sitting. I had seen a nest hatch when I was a young girl. The mound had bubbled like a volcano, the sand shifting with the little sea turtles below. Like an eruption, the hatchlings poured out of the ground. What began as a slow trickle, with only one or two, slowly became a steady lava flow until all 120 hatchlings broke free of their underground nest. This real-life Nat Geo moment was the experience that got me interested in working with turtles to begin with. After the brief explanation of the internship, Dr. Keely asked me the traditional interview questions.

When she said, "Tell me a little about yourself," I reiterated the standard few facts I had practiced with my parents on the phone. She inquired if I had reliable transportation to and from the hospital since it was about an hour's drive. She asked what experience, if any, I had working in rehabilitation or with sea turtles—I totally flopped on that question, seeing as I had absolutely zero. Interviews are supposed to feel like a professional conversation—there should be a flow between employers and future employees— but I remember the long awkward pauses between questions and anxiously wondering what Dr. Keely was writing down on her yellow notepad.

She ended with the stereotypical last question: "Why do you want to work with this organization specifically?" I smiled. I had practiced for this by looking up the rehabilitation mission statement. "I believe participating in the rescue, rehab, and release of these magnificent creatures will be a privilege. I look forward to participating in outreach and education of the local and visiting public while building a more thorough knowledge of rehabilitation methodologies through working hands-on with these animals myself." Dr. Keely smiled and nodded at my response and wrote an extra note in her pad.

While I had been totally unconfident until that point, that was when I finally felt a spark of hope. I left her office feeling okay and met up with my friend Kimmy, who had shown me the internship opportunity in our ArcGIS class and even helped me write the cover letter for the application. "I wouldn't worry too much about it," she said. "If I got accepted, you'll definitely get in, too." I didn't really believe her until a few days later, when an acceptance email showed up in my inbox.

The next day, we arrived at the facility. It was pretty unassuming from the outside. The Sea Turtle Hospital was a reasonably-sized single-story concrete building surrounded by a gravel parking lot, complete with a small sign to let visitors know they'd arrived in the right place to see some sea turtles. We punched in our codes that our mentor, Tina, had provided us and we opened the back door.

When most people think of sea turtles, they picture their characteristic shape, their iconic shells, and their big, beautiful eyes. What most people, like myself, don't consider is the smell. Fun fact: sea turtles stink. Not in a mild "Oh, that doesn't smell great" kind of way, but in a "Wow, some rotten eggs got left in the trash too long" kind of way.

We meandered our way through the back of the hospital to a small meeting room, not much bigger than Dr. Keely's office. It had a small cubby shelf, chairs lining both sides of the walls, and a fridge and a microwave at the back of the room. My fellow colleagues Tina, Karen, and Lindsey stood in the small room with me. We waited for the rest of the interns to arrive, filling in a couple at a time. At 6:50 a.m., Tina, the brown-haired woman whom I guessed to be in her forties, started talking. "Welcome, everyone, to the Sea Turtle Hospital internship! We are so thankful to have you here with us to help us run this hospital this semester."

After the introductions between the internship team and rehabilitation staff, we went on a tour of the facility. There were three rooms in the hospital, split into five main compartments. We began with the main hallway, which was the starting point for tour groups who walked down the long passageway learning of turtle facts and threats. These included the skeletal structure of the sea turtle, how their nest is made, how many hatchlings normally come from a nest, and the various threats that sea turtles face.

There are seven different species of sea turtles: loggerhead, green, hawksbill, Kemp's ridley, olive ridley, flatback, and leatherback. All of these turtles are classified as endangered species. Since we were based in North Carolina, the Sea Turtle Hospital mainly worked with greens, loggerheads, Kemp's ridley, and the very rare leatherback—I think throughout the entire history of the hospital, they had only received a handful. I will spoil it for you now—I didn't work with a leatherback. I did, however, work with a very spunky Kemp's ridley.

During our internship, we did a lot of cleaning, a lot of fish gut chopping, and a lot of poop scooping. That didn't bother me, though; frankly, I was just excited to be around sea turtles. Over time, I learned their individual likes, dislikes, and attitudes. Seeing them improve after coming in so fragile was heartwarming. Most of the sea turtles we treated were flown to us from the coast of New England. The Sea Turtle Hospital mainly received turtles due to an illness called cold-stun. Cold-stun occurs when temperatures drop below the 50-degrees-Fahrenheit range and sea turtles become incredibly lethargic. They will not eat, they will not dive, and they hardly have the energy to swim. They end up floating to the surface: a sea turtle ice cube. These turtles, carried by wind and currents, will often wash ashore on beaches and be rescued.

> "In North Carolina, winter weather events causing sea surface temperatures to acutely drop below approximately 12 degrees C [53 degrees F] may result in strandings of threatened loggerhead *Caretta caretta*, green *Chelonia mydas*, and endangered Kemp's ridley *Lepidochelys kempii* sea turtles due to cold-stun syndrome."
>
> (Niemuth et al., 2020)

That winter, we received over 100 cold-stunned turtles no bigger than the average-sized dinner plate, and that is how I met HoneyBee. HoneyBee was initially a cold-stunned Kemp's ridley sea turtle.

> "Kemp's ridley turtle (*Lepidochelys kempii*) is the smallest and rarest species of sea turtle. It is listed as critically endangered by the World Conservation Union and as endangered by the US Endangered Species Act."
>
> (Keller et al., 2012)

Well, HoneyBee came in critical condition. She was severely underweight and was so frail that we had to keep her water down to a level where she didn't have to pick up her head to breathe. This small, weak little turtle became the entire focus of my internship.

Sea turtles are reptiles and cannot regulate their temperatures like other marine species can, so they become at risk during winter months. Cold-stunned turtles are often juveniles, having not left the area before the cold of winter. Sea turtles will generally migrate to warmer waters during the winter.

> "They do this by swimming away from [the] shore to deep water or by migrating south. If water temperatures fall below 50 degrees F (10 degrees C) and turtles are present in the area, they are at great risk of becoming cold-stunned."
>
> (National Oceanic and Atmospheric Administration, n.d.)

When sea turtles become cold-stunned, they cannot use their flippers. This then impedes their ability to evade predators and seek food. If the condition continues, it could also cause secondary health risks to the turtle's life.

At eight in the morning, after opening tasks, we prepped food for our reptile residents. What I didn't know is that these turtles were very specific with their food options. Laid out on the chopping block were foods ranging from heads of lettuce to fish, blue crabs, and squid. The thawing out of these frozen foods contributed to the lovely sea turtle hospital smell that is oh-so-strong. Each species of turtle got different options according to the diet they would naturally seek in the wild. Loggerhead turtles got fish and squid, while green sea turtles were given more lettuce and fish.

There was one resident that was the exception to this: Lennie. She was another Kemp's ridley, found near Topsail Beach in North Carolina. She had been beaten in the head, causing her to go permanently blind. After being deemed non-releasable by the state, she became a permanent resident of the hospital in 2006. Lennie was the pickiest eater of them all. She rejected any and all food options but squid. No fish, no crabs, just squid. If you were the lucky intern prepping the food for her, you had to be more specific because of that. Lennie didn't *just* want squid; she wanted squid heads. No slimy tentacles for this turtle.

In addition to her food choice, due to her blindness, you had to use forceps to feed her. So, with my small bin of squid heads, forceps, and an iron stomach, I headed to Lennie's tank—a vibrant blue against the custom-made cement floor. Lennie lay gently on the bottom of her enclosure until I tapped on the side of it; this let her know that someone was there. Like a queen rising from her throne, she gently pushed herself off the bottom and floated to the surface. I leaned more than half of my body over the walls, and using the forceps, grabbed a perfectly-prepared squid head and submerged it into the water, watching the oils mix on the water's surface as Lennie made her way to me. I gently wafted the squid head back and forth near her head, pushing the smell to her nostrils. She caught the scent and, a little clumsily, took the squid from the forceps. This process repeated until she got all the squid heads and one extra, just because she was a special girl and I didn't mind cutting up extra partially-thawed smelly squid.

After being fed, each turtle received a scrub-down of their shell, or carapace (keh-ruh-pays). We did this to prevent any algae buildup and maintain cleanliness. This is when my intern group and I had the most fun. Equipped with industrial floor brooms, we set about our carapace-scrubbing duties. Lennie *loved* getting her scrubbings. She would actively wiggle her entire body back and forth along the bristles like a cat rubbing itself along your legs or a bear hitting just the right spot while scratching itself on a tree. Seeing her wiggles made the entire main wing light up with smiles and laughter.

The hospital's three main rooms began with the laundry room/kitchen, where we handled food prep and dishwashing. Second was the critical wing, where the sickest turtles were located. These were the turtles that needed more intense, detailed care—ones like HoneyBee. Lastly, there was the main wing, which housed the turtles that were out of critical condition but still needed to improve—most of them needed to gain weight. Alternatively, some turtles in the main wing were completely rehabilitated and just waiting on clearance from our vet and a release date.

Every morning, after the turtles in the main wing had been taken care of, I went to HoneyBee's tank. She lay in the bottom of the dark bin, her eyes closed, looking more like a turtle statue than a living animal. We lifted her and the others gently out of their tanks and washed them. Afterward, we washed the tanks and refilled them with new water. Unfortunately, these turtles were in such critical condition that they were on antibiotics or had additional health problems which caused them to have diarrhea.

Accompanying HoneyBee in the critical wing were turtles with various conditions. One was an adult female loggerhead that had been struck by a boat and had a cracked carapace. There were also about fifteen other juvenile greens that were severely underweight from cold-stun and refusing to eat—all of these turtles were so exhausted that they could not have normal water levels in their tanks for fear of them potentially drowning. The critical wing had a thick extra layer of stink, but it was all worth it the day HoneyBee finally opened her eyes.

She had been there two weeks by the time those big saucers looked up at me, and in that moment I could tell she was a fighter. My reflection shone in them as she picked up her head and her throat expanded, taking a heavy breath that seemed to say, "Can you not bother me today?" I grabbed Tina to notify her that HoneyBee had finally opened her eyes, and we proceeded to bother her. I gently lifted her serving-platter-sized body out of the tub for her daily bath. Unlike previous washing days, she started aggressively smacking her flippers, fighting me. She landed some painful slaps on my forearms, turning them bright pink, along with scratches from her small claws. Getting beaten by sea turtles was not on my expected list of things to happen.

Since she was feeling spunky, we offered her food. She proceeded to ignore the oily fish swirling around her still-shallow tank. Figuring she might be a shy eater, I walked away, but she still ignored it after several hours. Tina suggested squid. Maybe she was like Lennie?

To say that HoneyBee liked squid was the understatement of the century. She charged at this floating piece of sustenance with an insurmountable amount of ferocity. Turtles have the ability to take off your fingers with their jaws, so this was dangerous business. While other turtles in the critical wing were slow to move and calm in their tanks, HoneyBee was easy to spot. During feeding time, she turned her tank into an active war zone complete with a wave pool, splashing water absolutely everywhere.

Since we didn't want to be covered in water and risk our appendages, we started feeding HoneyBee basketball style. We stood five feet away and tossed the pieces of squid into the tank, laughing as she dive-bombed each one, throwing water all over the cement floor. Although we started to see that HoneyBee's name didn't quite match her personality, she was still weak and would tire easily. After feeding times, she seemed to settle, not stirring much after feeding took all the extra energy she had. After additional weeks of bathing and feeding, she was able to keep her energy levels up, consistently spinning around in her small tub. After one final checkup with the vet, it was decided that HoneyBee was cleared for the main wing!

Gathered together and rolling a cart, we moved her from the critical wing to the main wing. She flapped and wriggled, showing us just how big and tough she was now, and made a huge splash as she was put in her upgraded tank. Naturally, in retaliation for this change that she didn't agree to, she proceeded to stop eating for a week. In the world of rehabilitation, sometimes it's one step forward and two steps back. Soon enough, though, HoneyBee was back to eating basketball style and splashing even more.

HoneyBee stayed at the hospital for a total of fifteen weeks and gained more than ten pounds during her stay. During mine, I gained a knowledge and appreciation of these creatures that I never thought I would. I also realized that while rehabilitation was noble work, it wasn't for me. I was simply too attached to these animals to do it day in and day out. I was hugely disappointed by this, as I was surrounded by friends that knew and had a grasp on their passion from the get-go. I remember hearing someone say, "Oh, I'm definitely going to be working with sharks." I even recall Kimmy saying, "Turtles are it for me. This is what I want to do for the rest of my life, or at least until my knees give out," with a laugh.

Sometimes, figuring out what you *don't* want to do is just as valuable as finding out what you *do* want to do. So, off I went to test run the next job.

❧❧❧❧

To help sea turtles globally, here are some options for easily implementable things you can do. Limit your plastic use. Sea turtles primarily feed on a variety of jellyfish species. Jellyfish don't have the capacity to propel themselves through the water, and float mostly near the surface. Plastic bags also do this. Sea turtles often will mistake plastic bags for their favorite snack. While some turtles can pass these bags naturally through their digestive system, this is not the case for many.

> "Up to 52% of all turtles have ingested some sort of debris."
>
> (Schuyler et al., 2015)

That's if they even make it to the water to begin with. Each year, mother turtles will build nests containing an average of 100 eggs on the same beaches they hatched on. These beaches have been under constant threat of development and changes, leaving many nesting mothers little room. However, that is not the main danger to the youngsters. Believe it or not, it is lights.

Sea turtle nests often hatch in the middle of the night or early morning, and hatchlings have evolved to follow light to guide them to the ocean. This phenomenon is known as hatchling disorientation or misorientation. Mistakenly following lights put up by humans, hatchlings may crawl the opposite direction of the water and even re-beach themselves after making it past the shoreline.

> "Sea turtles emerge from the ocean to lay their eggs on coastal sandy beaches and are primarily affected by artificial lighting in two ways: first, artificial lighting can act as a repellent to nesting female turtles, affecting the density and arrangement of nests across developed beaches. Second, light pollution can disrupt the seaward orientation of hatchling turtles after they emerge and begin to move away from the nest, often resulting in fewer individuals reaching the ocean and lowering hatchling survival."
>
> (Price et al., 2018)

One of the easiest and most beneficial things you can do for your local turtle population is putting timers on your outdoor lights. Turning them off between 10:00 p.m. to 7:00 a.m. during nesting season, which is usually from March to October, can save multiple lives.

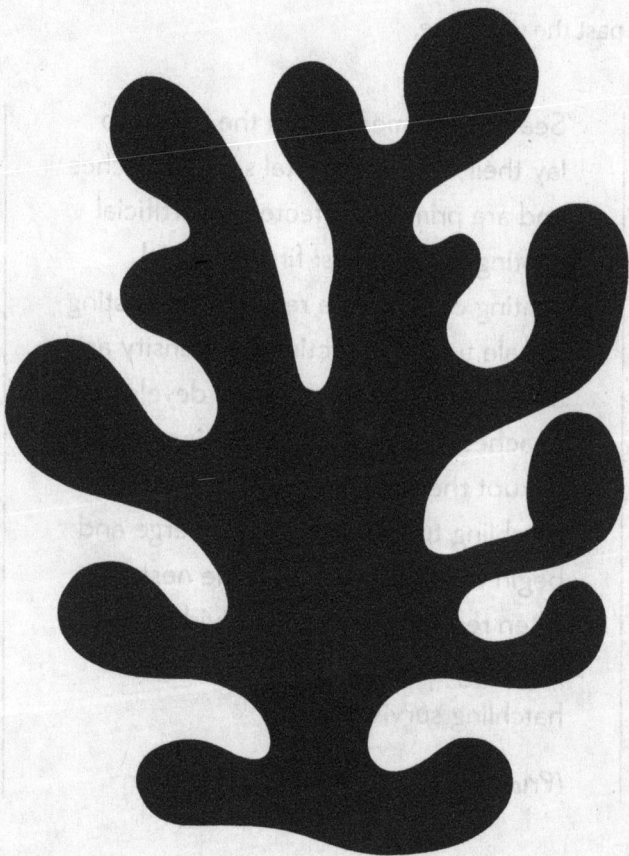

Chapter 5

Just Say Yes

During my undergrad years, I also had another door open for me that I said "Yes!" to. Bonnie Monteleone was a professor at my university and leading a research course on ocean plastic debris. Bonnie, or BonBon, is an icon in the realm of recycling and ocean debris research. She has conducted research in five gyres (large systems of rotating ocean currents) of the ocean, the Caribbean, and has a traveling art exhibit aimed at educating the public about the plastic problems happening in our oceans. Notably, she was also part of the documentary *A Plastic Ocean*. Bonnie's life mission is to make a difference by raising awareness of marine microplastics and ocean pollution as a whole.

During the beginning of her course, we watched *A Plastic Ocean*, which included researchers traveling the globe to highlight different perspectives on the problem of plastic in our oceans. One particular topic that interested me was the microplastic ingestion occurring in a variety of ocean animals, from seabirds to sea turtles and even large fisheries species.

Microplastics are defined as any plastic material that is invisible to the naked eye. These plastics are made up of a variety of materials, like Polystyrene (Styrofoam) and PVC, to name some common ones. Because plastic materials are cheap and easy to use, they can be found in food packaging, fishing gear, skincare products, clothing, and much, much more. Through the millions of pounds of plastic that are introduced into our oceans every year, microplastics are created when these larger pieces of plastic begin to deteriorate and break apart in the water column: the ocean between the surface and the seafloor.

> "Up to ten million tons of plastic enter annually in the oceans."
>
> (Almroth & Eggert, 2019)

> "This oceanic 'soup' of plastic is composed of different particle sizes: macroplastics (>250 mm), mesoplastics (1–25 mm), microplastics (MPs) (1–1,000 μm), and nanoplastics (NPs) (<1 μm)."
>
> (Hartmann et al., 2019)

One major concern is the effect of these plastic pollutants throughout the entirety of the food chain, known as trophic transfer. For example, a shrimp eats a piece of microplastic stuck in a piece of algae, then a crab eats the shrimp, then a squid eats the crab, then a yellowfin tuna eats the squid, and then we eat the tuna.

> "Microplastics (MPs) are an everyday part of life and are now ubiquitous in the environment. Crucially, MPs have not just been found within the environment, but also within human bodies, including the blood."
>
> (Leonard et al., 2024)

Part of the reason this is happening is the fishing industry, and human ingestion of fish with these plastics in their tissues. This phenomenon is impacting more than just the fishing industry, however—it's impacting the ocean as a whole.

> "Microplastic contamination has been documented in over 325 species of benthic and pelagic marine organisms caught across the world's oceans."
>
> (Miller et al., 2023)

This has further implications than plastic in your seafood entrée. Over the course of several months, I analyzed the gut contents of trophy fish species from the Bahamas with two other students. These fish were caught and donated to our research laboratory through the kindness of local fishermen in partnership with Bonnie. In total, 143 fish were dissected and analyzed. Most of these fish were mahi-mahi, yellowfin tuna, barracuda, and wahoo. They were frozen upon shipment to the lab and then thawed for dissection. Talk about a smell. *Please send me back to the turtles*, I thought. More than twenty deceased fish lay out on lab tables thawing for several hours in order to be dissected. Their stomachs were analyzed by the naked eye of the dissector, and then the contents of the stomachs were placed into a 10% potassium hydroxide (KOH) solution. This allowed the organic compounds to be dissolved, leaving only inorganic compounds behind (i.e. plastics).

In the middle of dissections, I decided that having a full-time job, going to school full-time, and doing one research project wasn't enough, so naturally, I added on an additional research project. I took on documenting samples of salvaged plastic pollution gathered off the coast of North Carolina and documented it in National Oceanic and Atmospheric Administration (NOAA) data collection reports. Samples were individually sifted through, counted, and categorized based on the type of plastic, color, and size. In total, there were over 1,000 individual pieces categorized both from nearshore and offshore samples.

Now, here comes the real-life science part: where things don't go according to plan. While the stomach content samples were labeled and waiting to be chemically analyzed, a large hurricane hit our lab building. It hit with such force that the roof was ripped off, rendering all of the samples compromised. So, all the data we had to analyze for that project was from the samples that were collected where plastics were found with the naked eye. Most of our samples did not apply, as most microplastics are less than five millimeters in size and impossible to see. But that's real science for you. Nine times out of ten, whenever you're doing any kind of research, no matter what it is, something is bound to go wrong. You'll have to modify your methodologies as you realize your original strategy is not the most efficient way of doing things, you'll have to change what systematics you're measuring to get quantifiable results, and all you can do is adapt and shift as needed. You can get all the way to the end of a project and realize that the components you were hoping to analyze were obtained the wrong way, your samples that were set aside at a different laboratory get lost in the mail, or a hurricane demolishes your building. It happens.

The reason that this topic is incredibly important is because plastic chemicals and pollutants can cause devastating effects in the body. It's no secret that plastic contains a plethora of chemicals—most are atrociously long compounds that are a mouthful. This includes "bisphenols, phthalates, polybrominated diphenyl ether, polychlorinated biphenyl ether, organotin, perfluorinated compounds, dioxins, polycyclic aromatic hydrocarbons, organic contaminants, and heavy metals, which are commonly used as additives in plastic production" (Ullah et al., 2023). Those chemicals really can't be good for the human body, an animal body, or frankly anybody, right? Shockingly enough, they aren't. These chemicals commonly impact the "endocrine components such as hypothalamus, pituitary, thyroid, adrenal, testes, and ovaries... Microplastics and nanoplastics disrupt hypothalamic-pituitary axes, including the hypothalamic-pituitary-thyroid/adrenal/ testicular/ovarian axis leading to oxidative stress, reproductive toxicity, neurotoxicity, cytotoxicity, developmental abnormalities, decreased sperm quality, and immunotoxicity" (Ullah et al., 2023).

Now, I'm not a doctor and I don't specialize in the medical field, but that's not good. That's not good for you, that's not good for me, that's not good for your pets, and that most certainly is not good for our ocean. The same ocean which supplies the majority of the global population with food, by the way.

So, what can you do? There are multiple large organizations now dedicated to reducing our plastic use and pollution. However, the largest shift can come from people at home. There are billions of people on the planet, and if we all shifted our plastic use, then that would generate an immeasurable impact. This can start with something as simple as your clothing, which can contain polymer threads. Try to buy 100% cotton, linen, and wool clothes. You can switch to reusable bags at the grocery store, including reusable plastic bags that can be washed, or even wax seals for storing food! You can use a fun, personalized water bottle instead of going through multiple plastic ones.

Frankly, I could fill the rest of this book with things you can do regarding your plastic use, because it's everywhere. Your clothes, shampoo bottles, sandwich bags, laundry detergent, takeout containers, phone, elements of your car—almost everything you own likely has some form of plastic in it. The best thing you can do? Look at your consumption and take a mental note of the things you have that are plastic. Ask these questions: Can you recycle it? Can you thrift instead of buying something brand-new? Can you reuse it? It goes without saying that these alternatives will take research and a bit of cost, but even if you choose only one of these alternatives, it's better than nothing at all. A large group of people doing something, even if imperfectly, is better than everybody doing nothing. Don't shy away from sustainable living due to the fear of perfectionism. Okay, tangent over.

ശശശശ

The last time I stepped foot on my university's campus, I was leaving to go on spring break. The email had read, "Due to concerns over the COVID-19 pandemic, we will be delaying the return of classes for an additional two weeks." I was sitting in my apartment when it was announced that the rest of the semester was canceled. I was sitting in my apartment when I got the call that my employer's restaurant had shut down and I was now out of a job. I was sitting in my apartment when I graduated college. I was sitting in my apartment when I applied for a new job at the grocery store. Just because the world stops doesn't mean that bills do, too.

I worked at that grocery store for eight months. I hated it. I was not working for the Earth. Instead, I was working for corporate America...arguably the polar opposite. The entire time I was working, I constantly searched online job boards. I only had one internship and a few research projects under my belt, so I figured I needed another internship, and Dr. Jack had agreed to talk when I emailed him for advice again. The problem was, with the pandemic happening, all of the internship programs were effectively shut down, with the exception of a rare few. I sent out five to ten applications, ranging from positions in Alaska to the Florida Keys. I waited weeks for a response, obsessively checking my email to break me out of corporate hell.

The Florida Keys coral lab was my top pick for a variety of reasons. I had been affiliated with the organization since I was a kid, it played a massive role in shaping my career choice as an adult, and after my experience in Belize, I was dying to work with coral again. They responded first. They also responded with a yes. I said *hell to the yeah* and got myself out of North Carolina.

The lab was a three-story cement building that was more reminiscent of a prison than an actual laboratory. The locals referred to it as such. When saying, "Oh, I work at the lab," they'd respond with, "That prison-lookin' place?" Yep...that's the one. However, this unassuming shell on the front end of the highway held the most glorious of secrets within. Of course they didn't want the building to be inviting; they didn't want people snooping around. Positioned on a canal, the glistening ocean was the backdrop to all of the scientists' hard work. There were rows and rows of tanks, down the entire back of the property, lined in neatly organized sections.

From what I could see upon arrival, holding my luggage and backpack, each was stocked full of coral. I walked up the stairs to hopefully find someone to direct me to the intern dorms. I knocked on a gray office door that matched the cement, and found the office of Sally, a stunning blonde woman not much older than me who was the office administrator. She quickly showed me the upstairs dorm room, where I snagged a bottom bunk. There were two rooms in the intern dorms, set up with two sets of bunk beds each. Inside were six other people who were interns for various departments at the lab. Each introduced themselves and the department that they were interning in. Most were like me, in the coral restoration department, while others worked in coral health and disease, ocean acidification, and coral reproduction.

I was paired up with another new intern named Kelly. She and I were starting on the same day, so we would have a built-in work buddy. The next morning, I met my boss, Zack. Zack had shoulder-length black hair and didn't look to be more than twenty-five. Despite his age, he was already the manager of the coral reef restoration program. He had worked there as an intern since the facility was even finished being built. He was the one that hired me as an intern. He was in charge of all of the interns as well as fellow staff of the program. On my first day, along with Kelly, he gave us a full tour of all the different programs onsite and introduced us to their respective staff in each laboratory.

The coral lab was made up of smaller labs that focused on different aspects of coral reef restoration. These labs were coral restoration (my program), ocean acidification, coral health and disease, coral resilience, the dive team, facilities, and education. Before becoming the coral lab, it was known as the "monkey lab." Rhesus monkeys were kept at a neighboring island—until there was some trouble. The monkeys developed a fondness for eating the red mangrove leaves and depleted the island; the county intervened and put a halt to the research project.

After my head was spinning from the sheer size of the facility and I started to blur the new names and faces, Zack led me back down to the rows of tanks in the Florida humidity and heat. The tank sections were labeled in alphabetical order with a number: A6, B1, C3, et cetera. In total, there were more than fifty tanks onsite. Every tank held an average of 1,000 or more coral fragments, depending on the size and species of coral.

Each coral was held on a ceramic plug and placed on an egg crate with a label to determine the specific genotype of the coral. Coral are animals, and quite aggressive ones at that. They will commonly fight each other using chemical warfare for space! This is known as "coral aggression." Therefore, each specific genotype of coral had to be separated accordingly to prevent damage.

Zack walked me through the main training protocols, handing me a printed sheet of different things to look for in the tanks, such as coral aggression, pests, discoloration, and an abundance of algae growth. He also showed me how to clean each tank and start a siphon using a piece of plastic tubing. Siphoning allowed us to clean the tanks without draining the entirety of the tank. Siphoning removes detritus, algae, and dirt that may end up in the tanks from various sources. Each tank also had an abundance of snails, roughly 200 in each, to help reduce the amount of algae growth occurring.

Zack then partnered me up with Katie, another intern who was soon to be leaving, to show me the ropes before rushing off to a meeting. Each intern was assigned a particular section of corals; my section was C. Due to the potential flooding from hurricanes, each tank was placed on a stack of three cement blocks on each corner, making the tanks chest-high. I worked my way through each tank with a trusty sponge, my siphon hose, and a step stool. By lunch time, I was entirely covered in salt water from both the tanks and sweat. It looked like I had just run a marathon through ninety-degree weather. I could not have been happier about it. I was finally working in my field, I was outside, and I was learning and developing new skills. Essentially, I was living my marine biologist dreams.

Over the course of the next thirteen weeks, I worked outside every day from seven in the morning until about four in the afternoon. I learned about each species of coral that we had onsite. I learned how to acid-wash tanks, wearing a gas mask and giant yellow gloves up to my elbows, sweat dripping down my face and rolling down my back. I moved entire coral tanks, scrubbed algae, and experienced the magic of the coral lab.

Chapter 6

Attack on All Sides

I 've written quite a bit about the various threats and different barriers that coral researchers have to overcome. So, what exactly are the threats that are plaguing corals and coral reefs? While there are various distinct reef types, coral reefs only make up 1% of the ocean floor, while supporting 25% of all marine life, and they're disappearing.

I ended up working at the Florida Reef tract. This tract is almost 350 miles long and starts in St. Lucie, Martin County, Florida, stretching to Dry Tortugas National Park outside of Key West. It is considered a barrier reef; however, the Florida Reef includes patch reefs. There are more than forty different species of corals there, and they are all in grave danger.

> "The Caribbean reef-building corals
> *Acropora palmata* and *Acropora
> cervicornis* have undergone widespread
> declines in the past two decades,
> leading to their designation as
> candidates for listing under the United
> States Endangered Species Act. The
> total area of live *A. palmata* at Looe Key
> is estimated to have declined by 93%
> and *A. cervicornis* by 98% during [a]
> seventeen-year interval."
>
> (Miller et al., 2002)

The quote above was published in 2002—over twenty years ago as of this writing. Can you imagine the further decline that has happened since then? It's devastating.

> "Almost 90% of the live corals that once
> dominated this reef system have been
> lost and coral cover is now estimated to
> be as low as 2–6%."
>
> (Donovan et al., 2020)

There are several causes for this decline, most notably: ocean acidification, coral diseases, habitat destruction, pollution, and coral bleaching events.

> "Coral bleaching, caused by the
> expulsion of symbiotic algae under
> extended thermal stress, has caused
> mass mortality worldwide since
> the 1980s."
>
> (Page et al., 2018)

> "Increasing ocean temperatures have,
> and will continue to stress coral reefs,
> and even if anthropogenic carbon is
> significantly reduced now our oceans
> will continue to be affected for decades
> to come."
>
> (Page et al., 2018)

See the theme here? We had all the bases covered at the coral lab, addressing each threat.

Corals came to us in the restoration program via three different routes; they could be wildly collected corals, bred in-house, or traded with a different laboratory.

The wildly collected corals came to us through construction projects, research projects, and a laundry list of permits. The assisted sexual reproduction corals were generated at the coral laboratory. Lastly, we did gene swaps with other laboratories to further diversify our genetic pool for restoration purposes. During every aspect of this work, all of these programs worked together.

The ocean acidification lab ran collections on monitoring acidity levels at each of the reefs we placed young coral in, a process known as outplanting, and maintained the Climate and Acidification Ocean Simulator (CAOS) system for coral resilience screening. Photogrammetry cataloged and organized all of the video and image data to create 3D models to analyze pre- and post-outplanting coral coverage. Coral health and disease maintained DNA samples to use for identifying genotypes and exposing different corals to disease strains. Education ran programs for the public to share our knowledge of corals and provide answers to both the "what" and "why" questions of our work. Coral reproduction made new baby coral and provided new genetic diversity to the reefs and the restoration team. The field team managed the offshore nurseries and reared a variety of different species that were better for in-situ nurseries (underwater structures in their natural habitat used to grow and cultivate coral fragments). Lastly, the coral restoration team tied everyone together by using both ex-situ (protecting species outside of their natural habitats, such as in nurseries, zoos, and labs) and in-situ nurseries, gaining new genets (genetically unique coral colonies) from the reproduction team, growing corals out, and then outplanting them onto the reef.

That's a lot. All of these problems are intertwined; they are linked to one another. Some you may have heard of and made your own assumptions about, and some are unexpected. Let's start with the unexpected threats to coral.

Algae is a huge problem regarding the safeguarding of these natural wonders. Algae are a diverse group of photosynthetic organisms that belong to the kingdom *Protista* (or *Plantae*, depending on classification systems). They are found in a variety of aquatic environments, including oceans, freshwater bodies, and even moist terrestrial habitats. Algae can come in a wide range of sizes and forms, from microscopic phytoplankton to large seaweeds. Normally, when most people think of algae, they imagine the green slime coating rocks along the sea shore, and they would be right! So, why exactly does this plant pose such a threat to coral?

The largest problem with this algae is the competition for space on the reef. Most corals grow at a rate of only a few centimeters per year, while some algae can grow that in a day. This causes the prime real estate on the reef to be taken up, preventing coral larvae from settling appropriately on the reef, and risks established colonies being suffocated by the algae growing over the coral. They also compete for light. When there is an abundance of algae present on the reef, this algae can create a thick blanket on top of the coral, blocking the sun and causing the coral to recede. Further research has uncovered:

> "The growth and survival of coral recruits was also severely reduced in the presence of macroalgae: survival was 79% lower in caged treatments and corals were up to 58% smaller with 75% fewer polyps. These data indicate that macroalgae has an additive effect on coral recruitment by reducing larval settlement and increasing recruit mortality. This research demonstrates that macroalgae can not only inhibit coral recruitment, but also potentially maintain dominance through a positive feedback system."
>
> (Webster et al., 2015)

This is further jeopardizing the delicate functionality of the reefs with breeding. Even if corals do make it to the settlement stage, there are no viable options on the reef to settle and attach, preventing them from entering the next stage, known as the recruitment stage. Hypothetically, if they do get a spot to settle, even with the competing macroalgae, there's also a potential disease to wipe them out. There are various identified diseases concerning coral reef researchers. Arguably the most pressing and prevalent would be stony coral tissue loss disease (SCTLD).

The concerning factor is:

> "This disease impacts twenty-four coral
> species in Florida, including several
> major reef-building species and those
> listed under the Endangered Species
> Act. The first report of SCTLD was
> recorded off Virginia Key, Florida
> in 2014."
>
> (Eaton et al, 2021)

Besides the ability of the disease to impact such a wide variety of species on the Florida Reef tract, it's also a very deadly disease, killing entire colonies in as little as a few days to a week.

To date, there are no existing diagnostic tools to positively identify SCTLD, making it difficult to determine if all impacted species are suffering from the same disease. However, there appears to be a hallmark ecology of SCTLD, specifically the variation in species susceptibility.

> "Highly susceptible species (typically first affected during an outbreak) demonstrate rapid disease progression, with total mortality ranging from one week for smaller colonies to one to two months for larger colonies."
>
> (Eaton et al., 2021)

Researchers are not sure if the disease is bacterial or viral. They believe it is bacterial, however, because it has shown response to antibiotics.

Currently, scientists are trying to stop the disease by using amoxicillin epoxy (a paste-like substance) and applying it directly onto the corals affected by the disease. During one of these applications, scientists noted:

> "Across five of the six tested species,
> the percentage of lesions halted
> using both amoxicillin bases was
> between 73-90%."
>
> (Neely et al., 2020)

While antibiotic treatments do provide hope, it's certainly not foolproof. Furthermore, with the current treatment plan laid out, it is nearly impossible to implement on a wide scale to cause significant difference. Each paste has to be placed by hand, monitored, and sometimes treated multiple times in succession. Nevertheless, researchers continue to strategize for monitoring and developing new treatment plans. That's if they can outpace ocean acidification.

Ocean acidification impacts any creature with a calcium carbonate structure. This group includes animals like crabs, lobsters, oysters, clams, and obviously corals. Ocean acidification doesn't just threaten these animals; it causes the entire breakdown of their skeletal base. As researchers noted:

> "Ocean acidification (OA) reduces the
> concentration of seawater carbonate
> ions that stony corals need to produce
> their calcium carbonate skeletons and
> is considered a significant threat to
> the functional integrity of coral reef
> ecosystems."
>
> (Guo et al., 2020)

Imagine trying to exist in the world, and then you find out oxygen is actually toxic and slowly dissolving your skeleton from the inside out. That's ocean acidification from the coral's perspective.

There are three calculated measurements needed to analyze ocean acidification: dissolved inorganic carbon, pH (how acidic or basic the water is), and total alkalinity. Keep in mind that when it comes to ocean acidification, we're talking about shifts in the entire chemical composition of the ocean. This is happening due to a variety of factors, but largely due to humans. To summarize simply:

> "Rising atmospheric carbon dioxide
> (CO_2), primarily from human fossil fuel
> combustion, reduces ocean pH and causes
> wholesale shifts in seawater carbonate
> chemistry."
>
> (Doney et al., 2009)

By using the latest technologies, coral researchers are hoping to help coral adapt to this.

At the coral lab, there is a CAOS system which, as I mentioned earlier, is the Climate and Acidification Ocean Simulator. Using this system, researchers are able to manipulate a multitude of ocean parameters such as pH, salinity, temperature, and so on. This system is used to assist in determining which coral genotypes will be best suited for predicted ocean climates in the future.

> "In order to mitigate the critical loss of
> ecosystem functioning, stress-tolerant
> corals need to be identified and
> protected concomitant with biological
> interventions aimed at maximizing coral
> growth, survival, and genetic diversity
> in degraded reef habitats."
>
> (Klepac et al., 2024)

In hopes of doing so faster without the large-scale and timely costs of running the CAOS system, the CBASS system (cyclic oligonucleotide-based antiphage signaling system) was developed.

> "An advantage of using CBASS over LT [long-term] experiments was the ability to determine nursery population thermal thresholds, which was 34.37 degrees C, and individual thresholds, which ranged from 34.00 to 34.72 degrees C."
>
> (Klepac et al., 2024)

The major win here for researchers is that while long-term studies take one to two months minimum, the CBASS system only takes eighteen hours to run, allowing researchers to get accurate *and* much faster results for coral resilience screening.

Speaking of temperature and determining which genotypes can survive in a higher thermal threshold, let's get into—trigger warning—climate change. I'm not going to get into an argument about whether or not we have global warming. We are seeing species extinction, disrupted weather patterns, and rising sea levels. Every fraction of a degree matters, which is why climate action is urgent, no matter what you believe is the political agenda behind it. Okay, phew, you survived! Now, what I will tell you is that yes, our planet has natural cooling and warming cycles. I will also tell you that over the past 100 years or so, we've had a drastically fast warming cycle that does not align with the planet's past warming cycle patterns. That being said, our water temperatures hit 101 degrees Fahrenheit off Key Largo in 2023. You cannot look me in the eye and tell me that it is normal for the ocean to be warmer than your hot tub.

Needless to say, the biggest threat to coral reefs is global warming. Despite all of the previous actions that I've mentioned in the book—all the ongoing research, policy change efforts, physical outplanting of new coral reefs, and protecting existing coral reefs—hundreds of thousands of people's jobs and efforts won't amount to anything if our environment can no longer sustain these ecosystems.

I know it's a brutal truth. Corals physically cannot sustain themselves with how rapidly our planet is changing. You're taking a million-year-old animal and forcing that animal to evolve in a short time span, which is next to impossible naturally. We're currently in the middle of our Fourth Mass Bleaching Event, which happens when large-scale marine heatwaves create severe coral bleaching over a wide area. These events are typically associated with high levels of coral mortality.

> "In 2005, the US lost half of its coral
> reefs in the Caribbean in one year due
> to a massive bleaching event. The warm
> waters centered around the northern
> Antilles near the Virgin Islands and
> Puerto Rico expanded southward.
> Comparison of satellite data from the
> previous twenty years confirmed that
> thermal stress from the 2005 event was
> greater than the previous twenty years
> combined."
>
> (National Oceanic and Atmospheric
> Administration, n.d.)

Actually, 2023 was the warmest year ever on record—
since global records began in 1850—by a wide margin.

> "It was 2.12 degrees F (1.18 degrees C)
> above the twentieth-century average
> of 57.0 degrees F (13.9 degrees C).
> It was 2.43 degrees F (1.35 degrees
> C) above the pre-industrial average
> (1850–1900)."
>
> (Lindsey & Dahlman, 2025)

To take the point further home, we're not only looking at mass bleaching events, but now categorizing them as mass extinction events, because the extended period of raised water temperatures not only causes the corals to bleach, but many to not recover. If we lose our corals, there would be devastating effects spread globally, affecting everyone on this planet, and certainly not for the better.

In my lifetime, I will never forget seeing firsthand the staghorn and elkhorn populations in the Florida Keys be wiped out. The majority of corals that survived and recovered from the bleaching event in 2023 are the bouldering species, and not the *Acroporidae* species. It was one of the most emotionally, mentally, and physically draining times of my life. "The marine heatwave in the summer of 2023 was the most severe on record for Florida's Coral Reef, with unprecedented water temperatures and cumulative thermal stress precipitating near 100% coral bleaching levels" (Neely et al., 2024), thus making restoration the alternative to losing our reefs.

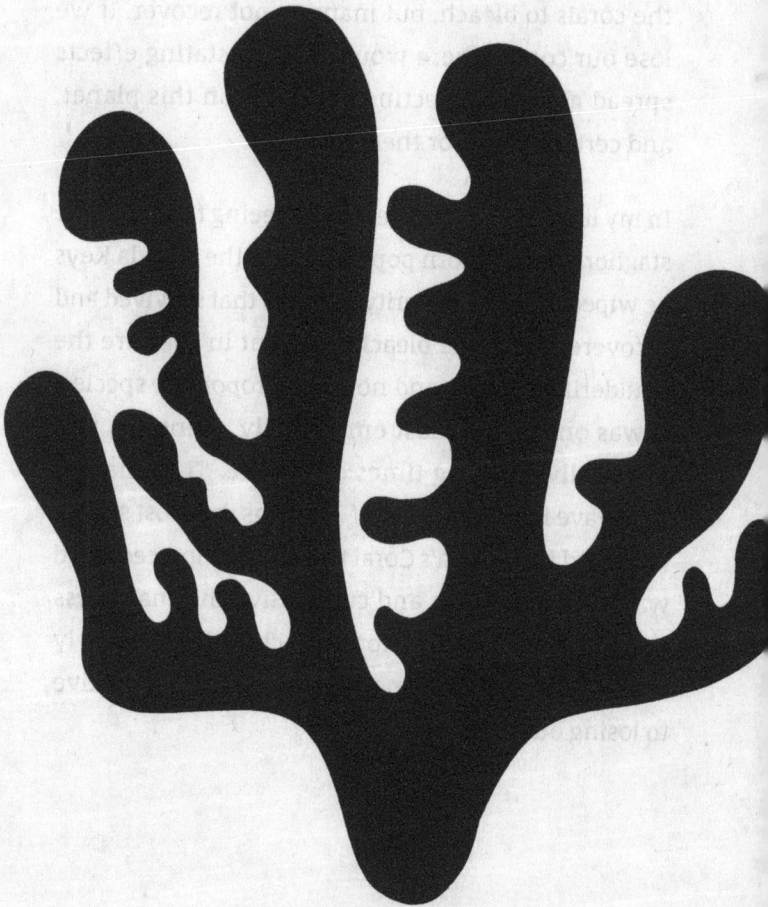

Chapter 7

Multifaceted Problems Equal Technically Advanced Answers

Corals face a multifaceted problem when it comes to their rapid decline. In turn, this requires scientists to have a multifaceted approach and understanding to solve such problems. By and large, climate change is the biggest threat to corals, but climate change is just an umbrella term to describe the various factors changing on our reefs. I've touched on the multiple problems our reefs face:

ocean acidification, algae growth, coral health and diseases, and lack of genetic pool—all side effects from climate change.

While scientists are working to improve our understanding of our environment and enhance restoration practices, the struggles with coral restoration are centered around the inability to regularly analyze data without physically being in the water. Collecting and analyzing data in the water is also constrained by variables such as dive location, boat availability, and the time available underwater when using scuba tanks.

So, how do we as scientists study coral reefs and restoration success without physically being in the water all of the time? Technology! We use different processes such as photomosaics, photogrammetry, Mars rover-like robots, and AI. Technology-integrated techniques in coral restoration are not anything new, but have rapidly become more popular.

> "In the last thirty years, the number of studies using photography and video as methodologies for data collection has drastically increased in marine science;

> species, communities, and habitats can
> be advantageously investigated through
> these techniques for different scopes."
>
> (Casoli et al., 2021)

Photomosaics form the basis of this process. Using overlapped and stitched images taken once by a diver, a process known as mosaicking, scientists can create one large photo map of a reef. This, then, is a tool that allows scientists to spend hours poring over different metrics of the reef, such as biodiversity counts, specific measurements of coral colony dispersion, and calculating size ranges of each colony. Rather than be constrained by spending minimal time physically in the water by comparison, this allows the painstaking review of these minor details.

Taking this process one step further is photogrammetry, which takes photomosaics and creates a 3D model from similar information. Photogrammetry = photographs + grammetry (recorded measurements). It pushes the data collection possibilities to new heights by not only measuring coral colony dispersion and size range, but specific size with height.

My role in photogrammetry was simply in the collection phase of the basis data. When restoration sites were completed with an outplant, marker buoys were placed on the perimeter of the site to indicate our stop and start points, as well as long rectangular markers to indicate scale once the videos were imput into data software. A rig was created using two GoPros fixed to a frame made of PVC pipe, timed to shoot at a specific frame per second. I know, very high-tech. However, it was built to be effective and handle field work. The scuba diver (a.k.a. me and other restoration team members) would then swim in a lawnmower pattern (all the way down one strip, then back the opposite way down the next strip) over the whole site, making sure to overlay the videos to allow for proper layering within the software. This is fairly straightforward, until you are in three-foot waves and attempting to swim in a straight line. Besides monitoring specific growth of coral colonies, this provides a unique database to see how coral reefs change over time, as well as to analyze effective restoration efforts, biomass and species assemblages, and even to quantify the loss of corals after a bleaching event.

Once these videos are put together and stitched appropriately, which is a several-weeks-long process, the result is a 3D model that can show the reef from any angle in detail. Besides our personal use for these 3D models, they also act as an educational tool for showing people restoration effort progressions throughout time. As noted by other scientists in the field:

"Currently, coral reef restoration is dominated by short-term and small-scale projects. While the restoration ecology community is aware of the need to increase both the spatial and temporal scope of restoration projects, there is no clear path forward on how to accomplish this, especially in difficult to access ecosystems such as coral reefs. Currently, photogrammetry is the best tool to achieve quantitative and standardized methods throughout the restoration process."

(Ferrari et al., 2021)

Besides photogrammetry, a plethora of technological tools have been developed, including remote sensing technologies like satellite imagery, unmanned aerial systems, and marine robots (Piñeros et al., 2024).

"Such methods provide information about the coral reef ecosystem, including its extent, reef type, geomorphological zonation, reef substrate, and benthic structure.

In recent years, drones and small autonomous surface vehicles have been used for mapping and monitoring coral reefs in shallow areas (less than twenty meters [in] depth). Furthermore, remotely operated vehicles have been used to monitor benthic habitats and marine fish communities, yielding good results as well as new challenges."

(Piñeros et al., 2024)

Recently, an AUV (autonomous underwater vehicle) was used to assist in a coral spawning event, described below:

"The presented robot design is reconfigurable, light weight, scalable, and easy to transport. Results from restoration deployments at Lizard Island demonstrate improved coral larvae release onto appropriate coral substrate, while also achieving 21.8 times more area coverage compared to manual methods."

(Mou et al., 2022)

Even in ex-situ systems, we had a startup from California approach the laboratory to build a scalable robot to use AI tracking systems to measure the growth of coral per day, organize every rack in each tank, and perform aquaculture duties. We even went as far as using lasers to remove algae from coral in our systems and have accurate data for planning outplants. Advancing technologies remove the limitations on physical work and allow for a wide database that can be shared and standardized globally. However, there are still many struggles to consider when it comes to the standard use of robotics in coral reef restoration.

Robotics and salt water don't always get along, especially in the "waterproof" department. Other issues include the maintenance of keeping these systems running without damaging the reef, accurately tracking where they are to retrieve them if necessary, and developing protocol systems of a seasoned diver to recognize and analyze their surroundings.

> "Robotic systems can aid in the automation and optimization of environmental variables, the detection of damage and changes in water quality, and the precise and efficient propagation and transplantation of corals. Furthermore, the information obtained by these systems can be used to improve conservation methods and provide crucial insights into the health and condition of coral reefs."
>
> (Cardenas et al., 2024)

The development of such technologies is an ever-expanding field, and though they're not fully automated, it is exciting to continue to leverage new tools to further individual scientists' capabilities for broader-scale impact.

Chapter 8

The Coral Restoration Program

W hen it comes to restoration tactics, there are some important distinguishing factors that play a role, mainly about the animal and environment we are trying to restore—you guessed it—corals. There are two main classifications when it comes to corals. The first is soft corals. These corals do not have a hard calcium carbonate skeleton. Instead, they have a tight interconnection of tissue that flows with the current. I would describe the texture like the root of a plant, almost fibrous. Additionally, polyps on a soft coral only have eight tentacles. Hard corals generate the calcium carbonate skeleton during their lifetime and are known to be the foundational structures for coral reefs. Hard corals are often known as "reef-building corals." Moreover, their polyps have six tentacles or replicates of six tentacles. There are thousands of different species of coral, and many, many more specific deviations of each. However, for the purpose of this book, you only need to know those two.

When it comes to coral restoration, scientists are primarily focusing on hard corals because of the foundational role they have in the development of coral reef ecosystems. Hard corals work like the foundation of a house. When you're building a house, you ideally want a nice, sturdy foundation to put the walls up. Without the foundation, everything will soon fall apart. Hard corals form the calcium carbonate structure for other organisms to settle on the reef that otherwise would not even begin to form! Because of the invaluable role that hard corals play in the formation of a reef, some reef restoration organizations primarily focus on those species. Some of the hard coral species that I worked with at my organization were:

- **ACER** (*Acropora cervicornis*), also known as staghorn coral

- **APAL** (*Acropora palmata*), also known as elkhorn coral

- **OFAV** (*Orbicella faveolata*), also known as mountainous star coral

- **MCAV** (*Montastraea cavernosa*), also known as great star coral

When it comes to these specific four species, there is one major distinction between them all—some are branching corals and some are boulder corals. These signifiers refer to their shape as a colony. Branching corals create branches off the main colony, while boulder corals look...well...like boulders.

The staghorn coral, ACER, as I'm going to be referring to it from now on, is a species that get their name from their antler-like appearance, similar to a stag. The branches can angle out in different directions, creating a web of interconnecting spindles of orange, light tan, and brown colors. They can grow at a rapid rate of several centimeters a year and are considered a critically endangered species. ACER is considered a branching coral.

The elkhorn coral, APAL, is similar to ACER coral, but in a much more dramatic sense. APAL are also named for their antler-like appearance, in reference to elk. However, instead of tightly interwoven branches, the branches of this species are significantly broader and larger. The chunky flat branches span out several feet from the base of the colony, growing to sizes of twelve feet or more. This species is also listed as critically endangered. APAL is considered a branching coral.

The mountainous star coral, OFAV, is a massive boulder colony, with polyps in little star shapes, hence the name. The color of the colony can vary from yellow to brown or green. This species can reach unbelievable size, with some colonies reaching twenty feet wide and over ten feet tall. OFAV is listed as endangered.

The great star coral, MCAV, is yet another boulder colony, with supersized polyps. These polyps also create a large star pattern across the surface of the colony. This colony usually develops colors in the brown, green, and orange families. This species can grow to monumental sizes, making themselves dominant characteristics on a reef. This species is also listed as endangered and, don't tell MCAV, but they're my favorite.

So, what exactly is being done when all these corals keep disappearing? We make more. In the land-based nursery, we use a technique called micro-fragmentation to create genetic clones of the corals. It's quite possibly the most outrageous thing you can imagine.

Now that we have covered all the corals, their characteristics, and issues, let's go back to my internship at the lab.

Zack led me upstairs to the restoration wetlab. I heard the saws going before I could see what was happening. "I'm going to teach you how to frag. You've been here for a little bit now, so it's time." He gestured to the saws on the counter. "These are called band saws. They're used primarily for jewelry making, but here we use them to micro-frag the coral."

These saws are simple, with one blade that spins between two discs. The blade is narrow, with one edge having a sandpaper-like feel to it. They are perfect for using in coral restoration because they cut comfortably through the skeletal base and the ceramic plug it's grown on. However, in case you slip, it will not cut your hands! It is not known exactly where the development of the micro-fragmentation technique originated; however, it has been around for several years prior to being used in restoration practices.

> "Fragmentation has been the method of choice to propagate branching corals for many years. Originally widespread in the aquarium hobby to produce and trade live Pacific corals, this method has also proven successful in restoring Acroporids on damaged reefs."
>
> (Page & Vaughan, 2014)

Zack walked me through the process of setting up the saw along with the few bins full of clean salt water on either side. One bin was for holding the coral that were to be cut, the next for washing freshly-cut pieces, and the last to hold the perfect new little micro-frags until they're glued onto a new ceramic plug.

I watched, mesmerized, as Zack demonstrated on the first coral and thoroughly explained each step of the fragging process. Scientists found that when corals are cut, their tissue begins to grow much faster than normal. This process is similar to how a human cut or wound heals at an accelerated rate. Corals are cut into the smallest pieces possible, roughly about one square centimeter, and in just a few months they can grow almost to the size of the plug. This is twenty-five to forty times faster than the normal growth rate for coral. Once these plugs are filled out, the coral is ready for outplanting back on the reef. Scientists do this by using a two-part epoxy and quite literally sticking them onto the reef. The corals will continue to grow outward and fuse together.

When considering the process of micro-fragmentation to a full-sized plug, then yielding an outplantable coral, there is very minimal wait time. From a freshly cut frag, APAL took two months to cover the plug, while MCAV and OFAV took significantly longer, with their timeframe being between six and nine months. We were careful to plant the same identical genotype together, because otherwise they will compete and not fuse together. The process of fusion between individual outplants in the wild in a cluster varies, but usually takes several months. There were outplant restoration sites from Miami all the way down to Key West.

Back in the wetlab, I gently tested the pedal on the floor to start up the saw. With a loud gnawing sound, the blade ran through the guides, and I gently placed my finger along the saw blade to reassure myself that it would not filet my finger off. Beside me, Zack chuckled at my hesitation, watching me with his arms crossed casually. Around us, the other interns were busy fragging away. I reached into the first bin and grabbed a large APAL plug, placed it against the blade, and made my first cut. Between every cut, I basted the coral with fresh salt water. Steadily, I repeated cut after cut until I ended up with fifteen little rectangles in the final bin. I placed the dot of glue on each stark-white plug and set the corals in their new tank. I could not wrap my head around the fact that cutting up these endangered species to make more was genuinely the answer. Talk about mad crazy science.

Soon enough, me and my fellow interns were fragging pros, generating a minimum of 1,000 corals each week. Before I knew it, I was quickly working through my days beginning with coral husbandry and ending with coral fragging. I was helping answer questions from the newer interns and had grown familiar with the coral restoration program staff. My internship days were coming to a close. I did not want to leave. At all.

So, naturally, I called my dad. "What do I do? I don't want to leave here; I don't feel ready to leave whatsoever." I stood out on the intern balcony, looking over the water.

"Well, you know, you can ask for a job?" My dad said, gently laughing on the phone. "You should go for it!" my mom chimed in, yelling in the background.

"That seems crazy, but they absolutely could use the help. There are only three staff members and they each have too much work to do," I replied.

"Then yeah, I would outline your points of why you're awesome, outline some of the points saying they need help, and just see what happens," my dad further encouraged me.

I drafted an email to my boss Elaine, a senior scientist and the vice president of research. In it, I briefly introduced myself and said I had been an intern for the past several weeks, I was trained, and that her team here needed some serious help. I wanted to be the one to provide that help. I read the email over and over again, scanning for any small grammatical or spelling errors, making sure it sounded nice yet professional. I steeled myself and brought it up to Zack. We were standing over a tank doing inventory for an upcoming outplant, about to finish the day.

"Okay, would I be insane if I asked Elaine for a job?" I blurted out. He looked up and raised his eyebrows, and I verbally repeated the email I drafted up for Elaine, making the argument that even if they didn't hire me, they should at least hire *somebody*.

"You've got some balls," Zack laughed.

"Those are not exactly reassuring words you want to hear from your mentor and potential future boss," I responded.

"She'll like that though; I think she'll even appreciate it. I don't think anyone has just straight-up asked her for a job before." He chuckled before shrugging his shoulders. "The worst that could happen is that they say no, right?"

"Yeah, yeah, just read the draft and make sure I didn't misspell anything?" I trade him for the clipboard he's holding and hand him my phone instead. I look over the inventory numbers and the chicken scratch notes made on the side of the printed out Excel spreadsheet. "The email looks good. Full send."

I sent the email that evening and waited. Monday morning, my bubbling excitement sent me nearly bursting into the coral restoration team's office. I didn't get the chance to say anything before Zack nodded his head, motioning a yes. "It's not a direct yes to you, but she forwarded your email to all of us. It seems like they're going to open up a position." Nobody could have managed to wipe the smile from my face, even if they desperately tried.

A few weeks passed, and nothing really happened. I needed the process to move faster; it felt like years were getting shaved off my life with the anticipation of any sort of news. Finally, Zack mentioned to me that there was a delay because some money needed to be moved around and allocated to cover the cost of the new person joining the team. After that, they would hold interviews to make sure that the chosen applicant pool was fair. "Just so you know, Elaine is coming to town in about a week, so she'll likely interview you herself when she's here."

When my moment came, I had just finished doing coral husbandry and had time to run upstairs and grab my notebook before I headed to a classroom to meet Elaine. I was headed to the very first real job interview of my life. I was also completely soaked in salt water, glue globs stuck to my shirt from fragging, my hair in a humidified bun. The epitome of a put-together professional candidate.

Elaine was a short blonde woman with half of her head shaved. She was no older than my own parents. She sat casually in the classroom with ocean-patterned leggings and flip-flops. She greeted me with a bright smile. Elaine talked like she had known me for years, and it immediately put my worries at ease. She asked me some of the more traditional interview questions, weaving them elegantly into the conversation. I answered in kind and tried to downplay just how nervous I was. I glanced down at my notebook I brought, scribbling little notes here and there of things she said. I prepped nineteen questions in total for Elaine at the interview. After reassuring that she had the time and didn't mind, I ended up asking her the majority of them. An hour later, Elaine still didn't give anything away in terms of whether I was actually going to get the job or not.

"That's the breakdown of the entire conversation," I recounted as I worriedly picked at my nails while sitting down with Derrick and Allie, staff biologists, and Zack at the lunch tables. "She didn't really give anything away, but she mentioned salary, so I'll take that as a good sign."

Derrick nodded his head. "Yes, I would absolutely take that as a good sign."

It would still be an additional two weeks before I heard anything for certain. I'm just going to mention: bless these scientists and their patience with me. I asked them nearly every single day about whether or not they knew anything, which they most undoubtedly did, and I wouldn't just shut up and be patient. Looking back, they made it pretty dang obvious, and I was too wrapped up in my own world to catch on. They gave me subtle hints of nodding yes, meaning *yes, you idiot, she's going to hire you!* And even a gift of a large stack of books from Allie, along with the note, *I'm sure you're going to have enough time to be around to read these and get back to me.* Yeah. It was that obvious. Mental note: don't be like me—take the hint and don't harass your friends, no matter how excited and nervous you are.

Two weeks after my interview with Elaine, I found myself talking with the team at the lunch tables. The weathered wood picnic tables had many functions— plug writing tables, an outdoor office, an impromptu meeting spot, and lunch spot.

Elaine walked up beside me, visiting from a different lab location for the week again. She gently put her hand on my shoulder and mentioned, cool as a cucumber, "So, you guys having an impromptu meeting with our newest staff member?"

My shocked look of surprise plastered onto my face as I whipped my head in her direction, noticing the big smile there. "Congrats," she said and meandered to a meeting down the hall.

That was just the beginning of me working as a coral restoration biologist in the Florida Keys for the next several years. I was twenty-two years old and had hit the career lottery.

Chapter 9

Intern Turned Mentor

After six months working as a technician, I was promoted to staff biologist. Jack replied to the news, "Ah, so the young student becomes the master," as I filled him in. Besides my duties working with the coral and the team, I was also assigned as the new intern coordinator for the program. Over the course of working at the coral lab, I was able to foster over thirty interns in their marine science journey. I never found myself gravitating toward educating and teaching others, but I did like talking to people. After all of the help that I received from my mentors, bosses, and professors, I thought of this experience as a great way to pay it all forward.

Working in marine science isn't for everyone. The great thing about internships is that they expose students to different career paths. I had interns that never wanted to leave while others couldn't wait for their internship to be over. Working in an ex-situ coral lab internship, you're constantly doing a lot of cleaning, organizing, and scrubbing. Add that to being covered in salt water, hot, and sweaty, and many people get swept up in the negatives. The important part is always directing your focus and remembering that everything you do, no matter how small, contributes to the bigger picture of the entire coral lab. Usually, our interns came in groups of two to five at a time. Some groups were shyer and more serious, while some were rambunctious and loud.

I worked my way down the stairs of the lab, headed to the tanks. Today was the newest internship group training day and the week ahead was going to be dedicated to showing them the ropes of the laboratory. I heard the giggling and laughter before I set my eyes on them. I smirked to myself; it was day one and I already knew this group was going to fall into the second category.

I did a brief introduction of myself and started the same tour that Zack walked me through on my first day. I found myself smiling easily, seeing their amazed faces walking through all of the programs at the coral lab. We finally made our way back down to the tanks and their excitement was palpable. Madeline,

Cailey, Sam, and Allie were nothing but smiles and fascination as we headed toward the first tank.

I stopped at the A3 tank and reached into the tank. The cool water came up my arm to nearly soak my shoulder. I picked a plug off the AP103 rack and pulled it out of the tank. They shuffled closer to me, eyes wide in astonishment. "Each one of these babies is destined to go out onto the reef at one point or another," I gestured to the rest of the tanks behind me. "Your job while you're here is going to be helping us take care of them and making sure they can fulfill their destiny." That was met with eager head nods and a "This is so freaking cool" from Madeline.

"Here are some important factors that you need to be aware of during your internship. One of the utmost is being organized." I flipped over the plug on the coral and pointed to the sharpie-marked label on the bottom of the plug, identifying this frag as the genotype AP103. "You'll see labels like this on every single one of these thousands of plugs, and labels on racks, too. Each letter coordinates to the species of coral that it is, and the following number is the specific genotype it is. AP refers to *Acropora palmata*, which is elkhorn coral. And then each specific number determines what individual it is."

I went on to explain that it works similarly to how the group of us are all human, but we each have unique DNA in comparison to each other. I further explained all of the different abbreviations of the various species onsite: AP, PS, PC, OF, MC, and so on.

I spent the day showing them different aspects of coral husbandry and the day-to-day schedule, learning about the interns as people, and answering various questions about different restoration tactics. The main points I always wanted to cover were not only how we were doing all of these things, but why we did things the way we did. Also, if they ever had any suggestions or questions, I encouraged them not to hesitate to ask anyone.

There are a wide variety of factors to assess and mull over when it comes to growing coral on land. Rearing corals in an ex-situ laboratory is not without its challenges. These corals are able to thrive in some of the most diverse and challenging locations on the planet, but growing them in a perfectly-lit climate-controlled tank—sometimes they decide they hate it.

Ex-situ, as I mentioned, refers to removing the subject from its natural environment and placing it in a controlled environment. In-situ refers to keeping the subject in its natural environment. Our coral lab had nurseries of both types. The ex-situ nursery was made up of tanks positioned at the back of the laboratory, outfitted with a shade structure, a flow-through water system, and temperature regulation. Our in-situ nurseries were made up of PVC coral trees anchored to the sandy bottom via rope several miles offshore. Our ex-situ corals were babied quite a bit by our staff and interns. Every day, I checked each one of the tanks, and I created a to-do list for the interns outside of their standard husbandry. Common tasks included additional cleanings of the tank, checking the water temperature within the tank, and fixing and switching out racks that were broken or overgrown with algae.

Every tank onsite was part of a flow-through water system, meaning that the salt water was pulled in via a pipe offshore, filtered, and brought to a regulated temperature prior to being pumped into our tanks. Every single tank had a constant amount of water flow supplied via the spigots that caused the tank to turn over every handful of hours. This ensured that our tanks were supplied with a constant stream of fresh salt water to maintain temperature and cleanliness. With the additional flow of water to the tanks, each tank was also outfitted with an underwater pump to mimic the waterflow that they would experience naturally on the reef.

To ensure our corals grew happy and healthy to fulfill their destiny of restoring the reef, there was an additional factor to consider in the ex-situ nursery: lighting. Corals are incredibly sensitive to light parameters. Too much light, and the coral will begin to pale and bleach out, which is our personal nightmare. Too little light, and the zooxanthellae in the coral tissue will not be able to photosynthesize correctly and the coral will begin to die off due to a lack of food. Over the entire outdoor nursery, a dark shade hung above our heads, providing necessary shade to the corals in such shallow water. Out on the reef, these corals could be anywhere between eight to twenty feet in depth, while onsite they were growing in tanks no more than eighteen inches deep at the most. Part of the daily checks for the coral was looking at all of the coral in the tank to make sure there was no discoloration occurring from unusual reactions to light.

These two parameters—water flow and light—are equally important. If either one goes out of whack, our corals don't do as well. Yet, there's a third additional challenge, and arguably the most time-consuming one. I have spent hours changing out racks, scrubbing out tanks, and brushing individual corals to defeat this menace: algae. Algae is particularly hard to maintain due to its incredibly fast growth rate, and with the nutrient-dense water in our tanks, it is a recipe for algae bloom galore. While some algae is normal in the tanks, the issue comes when there is an abundance. Algae blooms can cause suffocation for the corals and obstruct their ability to obtain the proper amount of light, potentially killing them.

Working alongside the interns, the other restoration staff and I taught them how to do all of these things: what to look for in the tanks as a red flag, the other research and outplanting parts of our jobs, and spending hours scrubbing algae off coral fragments. While we did all of these things, we also had to organize every single coral frag in the laboratory. We had to know what species a coral was, what specific genotype it was, when this genotype arrived at the lab, how many of that specific genotype we had at the lab at any given time, what locations that specific genotype had been outplanted to, and where they were all located throughout the various tanks onsite.

The key to maintaining all of this? High levels of organization and impeccably-kept records. Any time corals moved, whether it be onsite between tanks, to a new laboratory, to the in-situ nursery, or outplanted, it had to be diligently recorded and monitored. To help us maintain that organization, each coral frag was labeled with its genotype and species. Next, corals of all the identical genotypes were grouped together via racks. These racks were made up of PVC legs and egg crate to allow water flow both above and below the coral. To further assist our organization, each one of these racks was labeled with a label plug that digitally categorized the coral. The final step was that each tank had a unique label as well. In each tank alone, there could be anywhere from two to twenty racks, depending on the size and purpose of the tank.

A common and perfectly reasonable sentence at the lab could sound something like this: "Hey Allie, B6 has a ton of CCA (a type of algae) growing on the bottom and algae growing on the plugs. I also notice rack 101, 203, and 455 all have broken legs on them. Can you transfer the tank over to B3 today? Oh, and please be sure to leave AP321 in B6. We need to use it for an experiment later; I think it's on rack 124." To anyone else, it probably sounded like we were speaking in code.

While we maintained the nursery, I found myself often having long conversations with interns as we worked. These conversations could cover topics from corals to career advice, life advice, and even relationship advice. Some of these conversations stuck with me and others I forgot within a week. However, one particular instance stuck with me more than others. Before breaking for lunch, I was doing the typical rundown for career advising with the group. During this particular part of the conversation, I would ask them questions about their internship in detail and guide them on how to explain the work they did effectively, thus helping them relate whatever position they were applying for to this experience they had gotten.

"Okay, last one and then we're off to lunch," I said, looking down at my watch. It was a particularly hot day as the Powerpuff Girls, as they had deemed themselves, and I sat collectively at the picnic table. "Explain to me what a coral is as if I am an alien that has never heard of or seen one before."

Sam was the first to perk up her head as she said very confidently and very seriously, "It's a kind of aquatic plant, everyone knows that."

"Oh my god, it is *not* a plant. Coral is an animal, what are you even saying right now?!" Allie, Madeline, and Cailey loudly exclaimed.

I dramatically slapped my hand against my forehead in response. "Sam, Sam, Sam, please don't tell me that for all of the time you've been here, you thought that you were working with plants."

Sam's dark features and light eyes looked over at me, as she dramatically recoiled from the loud outburst of her fellow interns. She sheepishly nodded her head. "Yes?" I sighed. "I have clearly failed you all," I said in a feigned woeful voice. "Once we get back from lunch, I'm showing you each and every way that a coral is most certainly not a plant."

Chapter 10

Living Rocks and Their Population Problem

"**H**ow exactly is a coral an animal?" is a question I get often on the tours that I lead around the coral lab. Coral is pretty spectacular at being a really unassuming creature. Most people consider coral to be plants or rocks, which I don't blame them for. Before the eighteenth century, corals were thought to be plants due to their superficial appearance of plant structures.

I mean, you have this thing that looks like it has branches, it grows with sunlight, and it spawns to reproduce. Sounds like a plant to me! Well, corals said, "Nope, try again, and don't even think about calling me a rock." The best way I can explain how coral is an animal is by talking to you about ants. As we all know, ants are insects. Ants are individual little bugs that have specific body parts that serve to keep them alive. Ants also form colonies, the huge mounds of dirt that show up in our yards and cause itchy bites on our feet in the summertime. These ants all work and live together to create a functioning, happy colony. There are vast, key differences, but corals are similar to an ant colony.

Coral polyps themselves are animals; they have little individual body parts that keep them alive, much like how the individual ant is still considered a bug. The polyps then build these huge extravagant colonies together, too! These polyps can vary in size from a few millimeters to a few centimeters. A coral polyp consists of a main cylindrical body, tentacles, a mouth, a gastrovascular cavity, mesenteries, and a skeleton. These polyps then go on and form a colony. Similar to ants, every polyp contributes to the entire colony.

Now, this is where species identification and the high quantity of variation come in. There can be colonies that are only a few polyps, like soft coral species, or there can be ones where there are thousands of polyps, like hard coral species. Frankly, it can get really confusing really quickly. Coral can have a diverse expression of genes causing different tissue patterns, structures, and polyp numbers. Corals have a mutualistic relationship with zooxanthellae, which reside in their tissue. The zooxanthellae provide the coral with food, oxygen, and essential nutrients like glucose and amino acids. Zooxanthellae produce these essential goods for the coral through photosynthesis. While corals rely heavily on their zooxanthellae to obtain vital nutrients, they are also capable of capturing food on their own.

Corals can use their tentacles to capture food such as zooplankton and other small organisms in the water column. This particularly happens at night. The tentacles are equipped with specialized stinging cells called nematocysts that stun their prey as they bring it to their mouths.

There are two different ways for corals to reproduce: sexually and asexually. Sexual reproduction of corals involves a yearly event called spawning. This is a synchronized release of gamete bundles, which contain both eggs and sperm, into the water column by colonies. This synchronicity is brought forth by chemical cues in the water such as temperatures, tides, and the cycle of the moon. These gamete bundles are released by multiple colonies on the reef and meet in the water column. Fertilized eggs become coral larvae. These planula larvae (the free-swimming larval stage of a coral) have the ability to swim using cilia (small hair-like structures) and are microscopic to the eye. Thanks to the ocean currents, the dispersal of these larvae across the ocean can cover considerable distances. These larvae then choose a settlement based on a few key factors. They receive settlement cues such as chemicals like the presence of coralline algae on the seafloor, or even auditory snaps and cracks of a healthy reef. Once the larvae settle in their new home, they undergo a metamorphosis to properly adhere to the reef. Once adhered properly, the first polyp is fully formed, and then begins creating the calcium carbonate skeleton. Then, the budding and growing process begins: the polyp duplicates itself over and over again until we have a full mature colony, at which time the process would cycle again.

However, this natural process can take a considerable amount of time. You see, corals become sexually mature at a certain size, not age. This average size in corals is around six inches in diameter, about the size of a dinner plate. To have this amount of growth, a coral can take anywhere from ten to seventy-five years just to reach sexual maturity—and the settlement and larval stages are some of the most dangerous times to be a coral.

> "For clonal organisms such as corals,
> the smallest size classes (e.g., including
> larvae, newly settled recruits, and small
> fragments) suffer the highest rates of
> mortality. Coral colonies above a size
> threshold shift resources from growth to
> sexual reproduction."
>
> (Forsman et al., 2015)

When the newly settled coral finally begins to grow on the reef, this is where asexual production comes in.

Budding is the process by which polyps clone themselves: they form new polyps from previously existing polyps. The parent polyp develops a small growth which eventually detaches and becomes a separate polyp, very similar to how cells in the body perform mitosis. Besides coral being absolutely badass to the point of cloning themselves, there's also fragmentation. Fragmentation occurs when a part of the colony breaks off due to environmental damage, like in the case of a hurricane. The broken-off fragment, known as a coral fragment, can land somewhere else on the reef and re-establish itself as a new colony. This process is especially seen with branching corals, such as staghorn corals. At the laboratory, this process is accelerated and monitored via micro-fragmentation.

Here's where it gets fun, though. We can manipulate *both* the sexual and asexual components of coral reproduction. You've already read about the micro-fragmentation technique, so here are the methodologies for spawning.

Unlike asexual reproduction, spawning generates entirely new genotypes of coral and is vital for the continuation of our reefs. The adaptive potential of our reefs relies solely upon new genetic diversities having selective traits.

> "If populations are no longer self-sustaining through natural sexual cycles, the benefits of sexual reproduction may be lost (e.g., replenishment of depleted adult populations, population recovery post-disturbance, gene flow, and increased genetic variation). As such, scientists and practitioners are stepping in to carry out coral sexual cycles in the laboratory to ensure the benefits of sexual reproduction are realized for restored populations."
>
> (Koch et al., 2022)

We are observing a near-total failure of natural coral reproduction success, a.k.a. the ability for corals to naturally reproduce in the wild, throughout the Florida Reef tract.

> "Both biotic and abiotic factors affect population replenishment success, and hence, anthropogenic influences such as pollution, sedimentation and climate change can negatively affect critical processes such as reproductive synchronization in spawning species, successful embryological development, appropriate site selection, settlement, metamorphosis and in the case of reef-building corals, acquisition of the required zooxanthellae partner."
>
> (Richmond et al., 2018)

Effects from these problems not only impact the spawning process but the entire life cycle of the coral itself. There are fewer mature colonies able to produce viable offspring, and more interruptions to generate successful synchronized spawning. Furthermore, less larvae are able to even settle themselves onto the reef due to other issues, like an abundance of macroalgae. Then the entire cycle continues year after year, generating only sparse adult colonies too far from each other to produce viable offspring. This means that our reefs are no longer self-sustaining and unable to repopulate themselves.

To adapt to this issue of declining spawning conditions for corals in the wild, scientists are turning to assisted sexual reproduction (ASR). During the entire spawning process, scientists can record observations and create a controlled stable setting in a laboratory. The coral reproduction lab maintained specific spawning nurseries where sexually mature colonies were kept, and where there was a gravidness (a.k.a. pregnancy) assessment to better determine the lunar cycle during which specific colonies spawn. Gravidness refers to the presence of gamete bundles inside of the coral. Gravidness is estimated depending on whether gamete bundles are pigmented inside of the skeleton; this means they are likely to spawn around the next full moon. Once specific adult colonies are selected to be spawning soon, these colonies are moved into large round tanks at the laboratory. The stage is set via stable water quality temperatures and ensuring that the "mood" is set for future coral spawning. Think of it as the maternity ward at a hospital.

For one of these events, I arrived at the coral reproduction lab at 8:30 p.m. The lab hosts spawning nights where scientists from other programs can come observe and assist with the work that they're doing. I looked like an old Western bandit, dressed in dark clothes, a hat, and a bandana covering my face. One of the downsides of working with endangered species is that you cannot wear any chemicals whatsoever. You're lucky if you get deodorant. This means no perfume, no sunscreen, and especially no bug spray. By the moonlight and my headlamp, I stalked through the swarms of mosquitoes, no-see-ums, and gnats gathered around the canal's edge. There, I found the coral restoration team surrounded by glowing red lights, already recording observations and prepping the tanks for spawning.

Each tank held two adult colonies of ACER coral. These two colonies were separated with a screen, so we didn't have accidental coral babies being made. On each side of the tank where the coral parent was, there was a label along with a few other tools like a clicker, extra headlamps, vials, and pipettes.

I greeted my colleague, Callie, and her interns, who had fully converted to being nocturnal animals. During the spawning season, they worked all night and slept all day. The week prior, they began doing observations of these corals to make sure no presumptive spawnings were accidentally missed. Each species of coral has a unique spawning window; however, most occur around the full moon in August. I ogled the different setups of the colonies in their tanks and felt the excitement start to bubble in my gut. I placed down my two energy drinks and offered a helping hand for whatever was needed to finish setting up—it's not unheard of to go home at four or five in the morning.

Callie, as always, had a bright, bubbly mood, placing bets on which colonies would likely spawn first, changing the string lights overhead to a dark red color, and playing her coral baby-making music over a Bluetooth speaker. This earned both laughs and cringes from the small group of ten coral nerds surrounding us. By 9:30 p.m., the ninety-degree weather and no bug spray started to press on our minds. Imagine hearing the high-pitched whines of bugs by your ears, knowing undoubtedly they were attacking you through your shirt.

"Alright! Everyone has their assigned colonies. Please make sure that your headlamps remain only on red so as not to interfere with the spawning process," Callie gently reminded everyone. Red has the longest wavelength of visible light on the spectrum, making the least amount of impact in light-sensitive situations.

Everyone slowly trained their eyes on their colony, hoping to see any sign of setting. Setting is a behavior in which coral gamete bundles bud out in the mouth of the polyp just before they're released. Once setting is observed, it means that the coral will spawn at any minute. I looked intently down at my ACER colony, thinking, *Come onnn, come onnn.*

Around ten at night, Lanie, an intern, was the first to shout, "Mine is setting!" Callie went over and double checked, leaning over the tank, nearly putting her face into the water. She suddenly called out to another reproduction intern, Todd. "Colony forty-three set at—" she glanced down at her watch, "10:02 p.m."

I switched over to Lanie's side of the tank, prepped with an additional pipette and selfishly wanting to get a better look myself. Like clockwork, the small gamete bundles started slowly bubbling up from the colony like small peach-colored balloons. Seeing spawning in person is nothing short of being in absolute awe. Once the colony begins the spawning process, gametes more rapidly release from the polyps covering the coral branches. Like incoming rain, what starts out as a slow trickle slowly becomes a full-on thunderstorm. "It almost looks like reverse snow, doesn't it?" I nudged Lanie with my shoulder.

She nodded her head in response but kept her eyes on the coral in front of us. "I don't want to look away. This is absolutely incredible. I never thought I would see something like this."

I smiled, looking down and reflecting on just how few people had seen this firsthand, or even knew how cool corals were. "Me neither. Being a coral nerd sure has its perks."

Soon, we were surrounded by movement as the interns and scientists yelled out to record that their colony had set, and repeated times back for recording. Each person found themselves leaning over the waist-high tanks, using a pipette to collect each individual gamete bundle and place them into their specifically labeled tubes. After an hour, almost every colony spawned, creating a huge amount of genetic material to manage, organize, and breed with.

The fertilization process happens inside the laboratory, offering everyone on spawning duty a reprieve from the bugs and the heat. In the coral reproduction lab, gametes are placed together in specifically labeled containers describing the cross. Using this process, the lab has achieved a 99% success rate! Furthermore, we can study mature colonies for desired traits. Some desired traits for coral would be heat tolerance, disease resistance, and extreme fertility. We run various experiments analyzing each colony for these traits. The reason why these traits are deemed the most valuable is because, in restoration, we are focusing on the aspects of coral that are the most vulnerable and important for restoration of the reefs as a whole. Therefore, heat tolerance in the face of rising ocean temperatures, resistance to various coral diseases popping up globally, and the ability to produce an abundance of offspring are vital for the continuation of reefs as an ecosystem.

Corals spawn their gamete bundles in a broad net, releasing them to the surface of the water. These gamete bundles contain both egg and sperm, so when these bundles clash and break apart on the ocean's surface, fertilization happens. When this process occurs in the laboratory, we can manage the fertilization of the corals by crossing them with other colonies of desired traits. By doing so, we can create a wide scope of genetic diversity, as well as end up with a coral colony that is heat tolerant and disease resistant.

Once these coral larvae have developed into the settlement stage, the corals are settled onto ceramic plugs. They're then thoughtfully and meticulously cared for during the next year and a half to three years, depending on the species. Once the coral reaches a full plug size, it is then pushed into the coral restoration pipeline, other laboratories, studies, and programs. This feeds the coral restoration program with brand-new genetic variations to alleviate the population problem we're seeing on the reefs.

> "In Florida's Coral Reef (FCR), reef-building coral cover has been reduced to ~2% over the last fifty years from a combination of ocean warming, disease, and storms, leaving most of these reef ecosystems functionally extinct. As a result, coral restoration programs have emerged with the aim to propagate and outplant tens of thousands of coral fragments annually, in an attempt to buffer continued reef degradation."
>
> (Klepac et al., 2024)

However, there is more than just the population problem to keep in mind, and this is just one brief piece of the things being done to restore our reefs.

During this entire process, coral restoration labs, coral health and disease labs, coral resilience labs, and ocean acidification labs are also continuously working by focusing on additional perspectives and aspects of restoration as a whole. Whereas many of the programs did their work primarily on land, restoration had the chance to experience the best of both worlds: the charm and difficulty of working on land *and* under the sea.

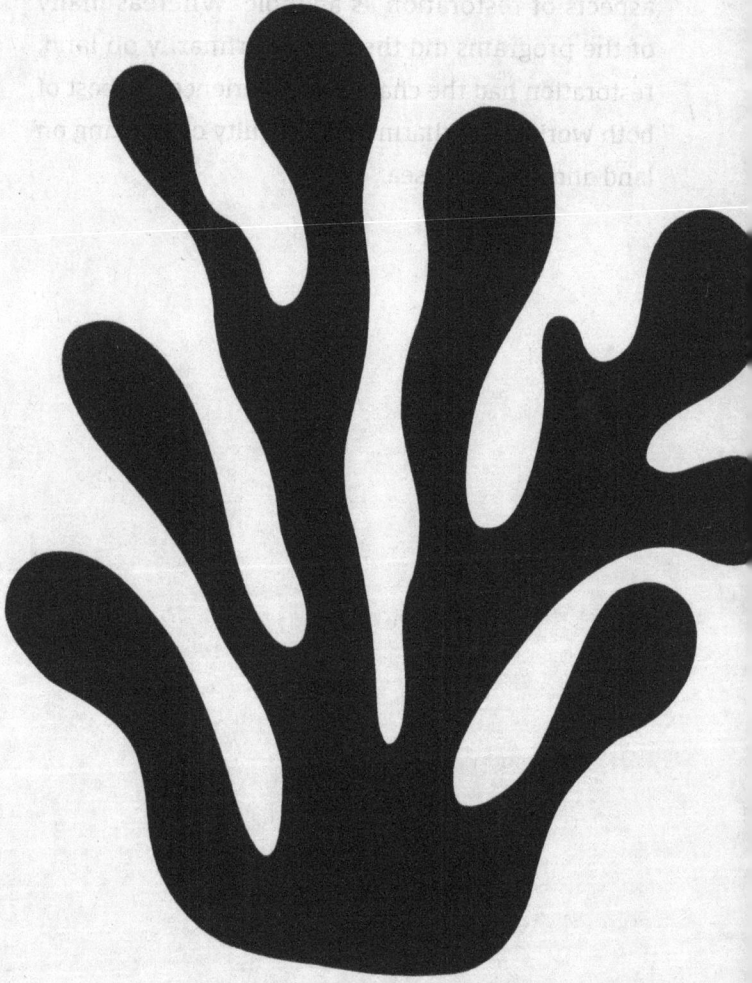

Chapter 11

Baptism by Fire

When I was hired onto the restoration team, I only had a whopping count of ten scuba dives and my open water certification under my belt. Allow me to underline, highlight, and italicize how absolutely absurd that is. Most of the people I was working with had dive numbers well into the hundreds, with the manager of the field team working as a professional diver since before I was even born. I had quite a bit of catching up to do.

"Are you sure you want me to go?" I checked my dive bag one more time, making sure I had everything I needed.

"You're going to have to learn sooner than later, and I would argue sooner is better than later, too," Zack replied. "Allie has things covered on land and this is outplanting—kind of integral to what we do, so you've got to learn. Baptism by fire."

The dive plan was set for everyone to meet at the lab at 6:30 a.m. the next morning. Having flashbacks to Belize, I prematurely packed the Dramamine, Bonine, and any other anti-nausea medication I could get my hands on. With the addition of a Dr. Pepper and some Goldfish, the iconic boat snack duo, my lunch was complete.

I got to the lab at exactly 6:30 a.m. and wanted to peel my face off my skull. I am not a morning person. I would much rather become a scientist of the night for repeated spawning activities than get up early. We packed the corals, now grown out from spawning, into large coolers. Instead of loading them with ice, beer, soda, and other beach snacks, we had about thirty eggcrate racks full of very mucus-y coral stacked and cushioned between damp bubble wrap. As long as the corals remained damp, they didn't need to be completely submerged in water for transport, especially when they were about to get outplanted.

We loaded up the truck with fifteen scuba tanks, our gear, and ourselves and drove to the dock. We pulled the truck as close as possible to the boat, walking down a ramp to reach our twenty-something-foot white Parker boat waiting in the water below. By this time, we were pushing nearly 7:45 a.m. just to get everything ready to go.

We huddled into the small seating area when Zack began to talk. "Okay, Derrick and I are going to be managing a lot of the heavy lifting and shifting the coral around since you're the newbie." It was Derrick, Zack, and Kyle, one of the field crew members, and I out on the boat. He proceeded to walk through the entire dive plan, showing me on a grid exactly where we would be and the outplant genotype list. It's vital to outplant corals of the exact same genotype next to each other so that they fuse. Coral fusion is when tissue merges into one uniform coral colony. Through this process, four individual outplanted frags become one coral colony, and are already closer to being sexually mature out on the reef. As a reminder, we're shooting for around dinner-plate size, as that's when corals become sexually mature to spawn. However, if those four fragments that are outplanted are not genetically identical, they will fight instead of fuse, thwarting the entire process. Using this outplanting technique, we can fast-forward time. For a coral to sexually reproduce, spawn, settle on the reef, and grow to roughly the same size, it could take anywhere from twenty-five to 100 years. We've shortened that to roughly five.

The boat swayed, the glistening turquoise of Florida Keys water surrounding us. Peering down into the water was like looking straight into an aquarium. You could see the individual coral heads and schools of fish moving in their own world underneath us. Before I could blink, Derrick was already in his gear and flipping off the side of the boat. I scrambled toward the back, uneasy on my sea legs and focused on not being queasy. I opened the back of the cooler, after peeling off the bubble wrap layers, and started handing Derrick stacks of corals. Zack soon joined him in the water, where they bobbed and shuttled to and from the boat like it was second nature.

Handing the last of the coral to Zack, Derrick said, "Alright, you see that marker?" He pointed to the bright orange kid arm floaty roughly seventy feet from the boat, barely visible over the rolling waves.

"Kind of?" I looked back down at him, bewildered.

"You'll meet us there. All the gear is down there and everything is moved. Just make sure you have your personal things," and, like a seal, he dipped below the waves and disappeared. It took me damn near fifteen minutes just to get situated with my personal gear. The boat was rolling, and my stomach along with it, making it difficult to find purchase on the flat surface of the boat to strap on a thirty-five-pound tank, a fifteen-pound weight belt, Calcutta snippers, a camera, and, finally, put on my fins.

Frustrated, I finally made it into the water. I had to surface twice and reorient myself prior to finding the marker. I struggled with my buoyancy and more than likely looked like a flopping, flailing duck rather than an actual diver. I was constantly adjusting the air in my buoyancy compensator (BC), unable to get my body to listen to me and act the correct way, and generally all the other super obvious characteristics that come with being a new diver. I had also already blown through about half my air after trying not to have a panic attack underwater from getting lost. Things were not going super well. I was so wrapped up in struggling that I could not appreciate the entire new world I entered. Zack glanced up at me, and signed, *Are you okay?* I motioned back *O-K.* What I meant was, *I am okay to struggle by myself and please do not analyze my horrible diving and let me get buried in the sand from my own embarrassment.*

Despite my trying to get the hang of scuba diving, I wasn't absolutely useless. I was able to actually get some work done. Carefully, I sunk my knees into the sand and began to open the epoxy containers and organize the tools from the tool bag. Each cluster on the reef was marked by silver aluminum tags. Each tag was hammered into the live rock, and had a number punched into it. Each number started with an X, indicating that these outplants came from ex-situ and were managed by the coral restoration program. These numbers are usually X-1 through X-500, up to even X-1,000. If it's a reef location that is used for multiple outplants, the tags coordinate with that so each individual outplant can be tracked. Along with that, each outplant site has a unique name.

When I first started working, each outplant only consisted of 500 individual corals; however, we were able to consistently bump that number up to 1,000. With an average of three outplants a month, that's 3,000 corals and 36,000 corals a year finding their way back home to the reefs. The largest-ever outplant done in one day was 2,000 corals!

Derrick glided over to the corals and gathered up the spreadsheet. Each tag mapped to a specific genotype of corals. For example, the genotype AP100 (APAL coral) coordinated tags X5 to X20. Swimming over to me, he grabbed a palm-sized amount of each type of epoxy. To adhere corals to the reef, we use a type of plumber's epoxy that hardens when mixed together. It doesn't dry out or rot easily, and it doesn't disintegrate underwater, making it a perfect material for us to use. It also has the consistency of warm peanut butter, making it easy to mix together. Derrick took off his gloves and mixed the epoxy expertly. With his ball of freshly-made epoxy in one hand and a tray of corals in the other, he swam over the crest of the reef out of sight. I set about the rest of my work and started cleaning tags. Unclipping the brush from my BC, I slowly started scrubbing the tags, removing built-up algae one by one. Oftentimes, outplant sites are prepped ahead of time; the site is surveyed and all the tags are laid down. This makes it so that when the outplant day comes around, we're able to solely focus on getting corals out onto the reef.

I glanced down at my air supply and shook my head. I was already nearing 600 PSI, meaning I was going to have to swap my tank out already. I managed to find Zack underwater and motioned that I had to swap out my tank. He nodded in acknowledgment that I would be gone and I began my ascent to look for the boat. It took me another twenty minutes to get back to the boat, swap my tank, and come back again. Zack and Derrick could do it in under ten. It was rough.

When I came back, I returned to my work of scrubbing tags, making it easier for them to find the cluster locations. My first ten or so dives for work were similar to this one: struggle a lot, organize the tools, scrub tags, try not to get too distracted by diving on a coral reef, rinse, and repeat.

"You just have to get some practice. Field work is not a strong suit for everyone, but once you get the hang of it, you'll feel a lot better," Zack reassured me in his office the next day. I continued to struggle and was not the best field person out there. There were many learning situations out on the boat. I would get sick, I would forget gear (whoops), and I would get nervous if I found myself alone. As I completed more dives, I learned how to overcome these challenges. I found a work-around to get my gear on faster, I discovered a loophole that allowed things to get done more easily, and I finally got my sea legs. With the help of my organization, I was able to get my rescue scuba dive certification. One particular field day, it all just kind of clicked.

I found myself the first to flip off the side of the boat. Megan, a colleague vital to the team who we all jokingly called Meggo, tossed me a heavy bag of epoxy and I checked my compass before slipping under the waves. I sank quickly, with the help of the additional weight of our tools. I made my way toward the outplant sight, periodically checking my compass. My personal tools—a clipboard, extra brushes, clippers, and my camera—gently bobbed on my carabiner clips against me, at this point a normal presence instead of overstimulating. I looked around, taking in the great big blue around me. A reef shark glided along, following the large crevices between the reef crests, and I smiled, thinking to myself, *How many people can say that the ocean is their office?* Sunrays poured onto the surface, creating a crackling of lights dancing on top of the live rock and coral, and making the surrounding fish glitter as they danced around the various sea fans, swaying through the surge like a gentle breeze. It was a fantastic day to be underwater.

I dropped the tool bag and made my way back to the boat to receive more gear when I passed Meggo and Zack, who were headed to the site themselves. After being handed coral, I gently made my final descent for the next hour. I flowed through the outplant protocols like clockwork—clean my tags, mix epoxy, grab my corals, and voilà, you have outplanted corals.

Meggo and I were usually able to finish our plots early and assisted the others with their corals or just plain old goofed off. We were able to carry on full-blown conversations underwater through hand signals and body language, float upside down, check out crevices for nudibranchs (mollusks without shells), and perfect our underwater photography skills.

Field days weren't always just work; they became fun, they became inspiring. Outplanting blossomed a deep sense of hope in my chest. We were doing something; we were on the cutting edge of science saving our reefs. Putting those little baby corals onto the reef after fragging them by hand and watching them grow in the tanks, coming back to the site a month later to monitor and record them beginning to fuse and start to branch, seeing hope grow on some barren rocks under the warm sun and surrounded by cool turquoise water—these were some of my favorite days at work.

Chapter 12

Coral World

O nce a year, our organization held a coral restoration workshop. Globally, there are many different workshops and conferences that scientists attend to share the latest research and ideas, and forge partnerships for future collaborations. They are held in the Florida Keys, such as the global symposium Reef Futures, or all the way on the other side of the globe, such as the International Coral Reef Symposium (ICRS), and many more locations. The importance of these conferences cannot be overstated. These chances for the top minds to come together to branch into the newest concepts of marine science is incredibly beneficial.

During a coral restoration workshop we held, over thirty different scientists from all over the world attended. These were individuals from the Caribbean, Belize, Hawaii, and even as far as Saudi Arabia. Despite the differences in our approaches to overall restoration practices, there was one universal thing that we could all agree on; we were at code red, with sirens, railroad warning signs, and fireworks going off for good measure with how much coral we were losing and at such a rapid rate.

We talked about the individual materials that we used, what had and hadn't worked for us, and asked each other questions. We walked attendees through our individual practices, from how we siphoned tanks to quarantine protocols for wildly rescued corals. Arguably, the debates and meetings were the most beneficial. There were many different questions to ponder during all of this, like: How can we increase the speed of this process? How do we outplant more corals on the reef without using more divers and more boats? Which is the better laboratory setup, in-situ or ex-situ? What are the pros and cons of each? How do we continue to handle the growing climate change problem that is causing widespread bleaching events? What are our bleaching coral protocols? How do we quantify and measure success when it comes to restoration? By the individual outplanted corals? By the cluster? By the corals that stay alive after a certain number of years? How do we manage coral genetics in the long term? How do we make sure we're not creating more genetic bottlenecks by planting corals that are related to each other? How would we even track that?

These are just some of the questions that we debated, and we are constantly trying to figure out how to do better and make the process easier. However, many focused on how to protect our coral restoration efforts and further adapt them to a rapidly changing climate.

Climate change, driven by greenhouse gas emissions from humans, is the most severe threat facing coral reef ecosystems. These emissions cause ocean warming, ocean acidification, and more frequent and intense tropical storms and heatwaves. Among these effects, ocean warming is the most severe threat to reefs; it causes coral bleaching, as discussed in Chapter 2 (Voolstra et al., 2023).

> "Mass coral bleaching has been increasing in frequency and intensity over the past decade(s) and caused a 30% decline in the global coral population. Recent estimations predict that, if global warming exceeds 1.5 degrees C, 70–90% of reef corals are at risk to be lost, and 99% will be lost if global warming exceeds 2 degrees C above pre- industrial temperatures."
>
> (Voolstra et al., 2023)

While there have been strides in the continued restoration of corals, there is not much that we can prevent or do when we grow them in a controlled environment, breed them, outplant them, and then... they die. So, what do we do? We're hitting the point now where there are no ideas that are too crazy. At the conference, there were wild proposals and ideas being spoken aloud.

"Why don't we glue small corals onto discs and dump them from a plane? This would cover a huge area and allow us to quadruple the outplanted corals onto the reef."

"Let's build a giant shade cloth net over the reef to reduce UV rays and prevent corals from bleaching out."

"How about using large offshore tubes to pump colder deep water nearshore to drop the temperature during heat waves?"

"What if we built a robot similar to the Mars rover that can monitor reefs without physical divers in the water?"

Absolutely no ideas were off the table. One idea enlisted the help of some geo-engineers and sounded like it came from a sci-fi movie:

"In March 2021, Australia's Great Barrier
Reef played host to a rare field test
of marine cloud brightening (MCB)
technology. Blasting a seawater mist
into the air from the aft of a research
vessel using a turbine outfitted with
novel spray nozzles, researchers
watched the resulting white plume rise
into the sky."

(O'Neill, 2022)

This effectively created huge plumes of salt water clouds to provide coverage for reefs and the lower temperatures that corals so desperately need. However, it came with some serious drawbacks:

" 'You could deploy SRM [solar radiation
modification] and then realize you
have changed the regional atmospheric
circulation and the country next door
suddenly suffers droughts. What is the
governance of this? Who is responsible
for the damage?' "

(Philip Stier as cited in O'Neill, 2022)

Alright, so maybe not the best idea, at least until more research is done and further technology is developed to accurately predict the possible repercussions of deploying marine cloud brightening. Let's take the crazy sci-fi theme to further heights:

"Electrified corals survive repeated severe bleaching events when over 90% of corals on nearby reefs die. Electric reefs are open mesh frameworks with more vertical levels of holes and surfaces than natural reefs, which can be grown in places where natural reefs cannot grow due to lack of substrate or unsuitable physical and chemical conditions.

They strongly enhance reef physical structure, wave absorption, ecological function, biodiversity, productivity, and habitat and ecosystem services including shore protection, sand generation, and fisheries habitat, even at severely degraded sites where no natural regeneration takes place."

(GOREAU, 2022)

By providing a low electric current, these mesh frameworks improve coral survival, and have proven to be a plausible answer to restoration problems. These "electric reefs" have largely been deployed overseas in areas like Indonesia. While technology and ideas get thrown around, and while scientists continue to further develop a variety of practices, coral restoration as a whole is still a developing field. Trial and error and "throw it at the wall and see what sticks" methodologies are still being deployed as researchers attempt to find the best ways to restore and aid long-term coral survival.

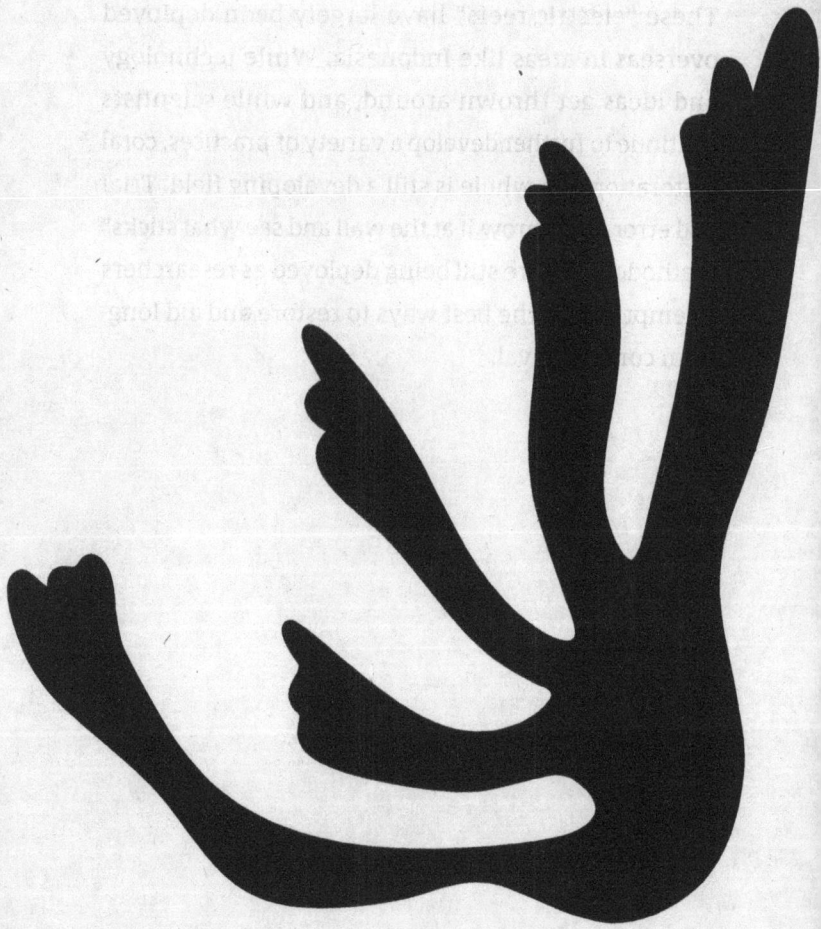

CHAPTER 15

Chapter 13

Hell Week

July 2023. The pit in my stomach felt heavy as I threw on my gear. I was just in my leggings and sun shirt; the water was so hot that you didn't need more than that, didn't want more than that. If you wore even a three-millimeter-thick wetsuit, it would feel like you were sweating underwater. I glanced at Meggo, her red hair glinting in the sunlight of the glaringly hot morning. We stared at each other for a long moment, the pit growing poisonous and curdling my stomach. We took a deep breath and nodded to each other, rolling backward into the hot tub that was below the boat. The bubbles from rolling in cleared from my mask as I looked down and something in me cracked.

Just a few weeks prior at this particular reef site, we outplanted over 1,000 corals. It had been a newer location but still decorated with fish and our baby outplant corals, peppering the distinct large pieces of live rock. We'd spent the entire day and then some getting these corals here, caring for them in our nursery, and finding them a good home for absolutely nothing. Those corals shining back at me were stark white against the live rock backdrop, blaring an alarm that something was very, very wrong. I took a shuddering breath and sank to the bottom, following my heart. Sinking. Down, down, down. My eyes welled in my mask as I took in the devastated site.

It wasn't just the corals. It was the sea urchins. Sea cucumbers. Fish. Crabs. Their carcasses decorated the graveyard I swam through. It felt like I was swimming through sludge. I paused, looking at my dive buddy. She was shaking her head and gave me an aggressive thumbs-down. We had to find something.

We split up across the site, cameras in hand, and began documenting. I floated from outplant to outplant, counting the aluminum tags as I went, searching between large rocks. Searching for anything that wasn't bleached. Clusters upon clusters of corals were bleached, dying, or fully dead. I fought down the burning in my throat, concentrating on the fact that I didn't have enough air to cry with the tank strapped to my back. I was twenty feet underwater; that wasn't the place to cry. I made it to tag ninety-two when the tears came. We found nothing. The entirety of the reef, the entirety of the outplant site—all of it—had been extensively bleached, with very few remaining outplants that were alive, their white transparent polyps waving in the current like a surrender flag. Surrendering to the heat.

We clambered back onto the boat. Out of the four of us, no one said anything. Meggo pushed her hair out of her face and piled her curls on top of her head, her expression tense. I looked at Derrick, the most experienced of us all, and the one who had seen the most shit.

I shrugged off my gear, staring into space, letting my mind crawl with all of the potential answers. I reached for my towel, wiping my eyes, hoping they didn't look too puffy. I grabbed my water bottle next to the captain's chair. Derrick's face was grim as he looked at his watch, always trying to keep us on schedule. I broke the silence. "Do you think there's a chance of recovery? I mean, have you seen anything like this?" I said it loud enough for the others to hear on the boat.

He turned to all of us. "This is completely unprecedented. I've never seen, and I doubt anyone has seen, anything like this outside of Australia. Yes, we've had bleach events before, but this is so severe." He shook his head, still wearing the grim expression on his face despite the donned sunglasses and visor. "I doubt anything that is alive here will be able to hold on before the water cools down enough." Meggo looked down at her hands, already running the math in her head, of the time frame and temperature it would take.

I worriedly questioned, "C'mon, we have to do something. We can't just let everything die?"

Derrick looked at me, and said, "Then pray for a hurricane." The boat ride was pretty quiet after that.

Zack met us at the truck as soon as we pulled up to the lab. He looked so tired. He had dark circles under his eyes, matching my own and everyone else's at the lab. No one was sleeping; everyone was barely eating. In greeting, he said, "They're calling another emergency meeting. I'm likely going to be on the phone well into the night, but I will communicate updates to you all." He glanced between us, likely looking at the same ragged expressions he wore, not even bothering to ask how bad the outplant site was—he already knew.

He did text throughout the night. My phone glowed with light every couple of hours. The last message was past midnight. It read, "We're evacuating everything." My mind jumbled at the thought. *Everything? The entire nursery? The thousands of corals?* The choice was either to leave them there and risk the water temperature not coming down fast enough, or risk potentially killing them in transport while trying to save everything. I tossed and turned, contemplating the pros and cons of each. How do you even make a decision like that? The stars had just started to fade when my mind finally fell silent. Exhausted, I pulled myself from the warmth of my bed. Evacuation plans. We had two main nurseries offshore, not even counting the mature breeding colonies for the reproduction lab. Fully stocked, they held thousands of fragments to replenish our reefs. Entire generations of different genotypes hanging by nylon strings on PVC trees, hanging on hope that they'd last long enough to get them out.

I meandered through my thoughts as I did a health check on site, pump function, and a general list of what was going on with the coral for the day. I wandered through the neatly organized rows of the tanks until it was time.

We met in the main office, big enough to hold the sizable team of twenty-plus scientists, divers, and managers. Our new manager, Richard, who stepped in to run the coral restoration program after Elaine was promoted, explained the plan to us. We were to evacuate the nurseries with our entire dive team operating underwater; our genetics scientists were to manage the catalog and organization of the specific genotypes and a driver was going to move them to our onshore laboratory, driving overnight to minimize the trip stress. It was daunting. It was also incredibly risky. That many corals in a shipping container could turn anaerobic (deprived of oxygen) and suffocate before getting to the secondary lab—not to mention the high likelihood of genetic mix-ups. Many didn't agree with it, but we could not come up with any other viable options either. A blanket of anxiety, stress, and uncertainty coated the entire lab like morning dew.

Our team in the secondary lab was too small to handle the influx of giant coral shipments by itself. Having someone from our lab would hopefully make the process easier. I looked at Zack directly and gently nodded my head. Zack was needed here, Derrick was a talented diver and boat captain, and our other teammates lived alone with their animals and were newer at the time. The logical decision was for me to go first. After the large meeting, the restoration team had a program meeting. I offered to be the first to go to the mainland lab and everyone agreed. The day proceeded to go along as normal, but everyone had the ghost of nervousness floating around them. There was no happy music; everyone talked in hushed tones and often ate their lunch while working.

"I'll keep you posted on when the go is. We don't have a timeline yet. We're building the plane as we're flying it, so to speak," Zack mentioned.

The phone call for go time came on the way home from seeing a movie with my husband the next evening. Richard spoke gruffly and directly. I would have a hotel room available for me near the secondary laboratory, I was expected to be there no later than tomorrow evening, and I could expect to be there for a week minimum. His words lacked the emotional sensitivity or acknowledgment that we were all in deep shit. My husband bristled at his tone, but I didn't have time to be worried about that. I didn't have time to deal with the sinking feelings— there was work to be done.

I arrived at the secondary lab by four o'clock in the evening the next day. I bypassed the hotel and drove straight to the lab instead. There, I found a six-person team. Some faces I gladly reunited with and others were new, but there was one distinct shift with this team. These people, blessed be, weren't drained yet. They were our reinforcements. They were happy, rested, and looked ready despite knowing the tsunami of work that was coming their way. I probably looked like a walking stress ball to them.

By the third day, we were running on fumes, loud music, and cases of energy drinks. The first shipment arrived sometime around three o'clock in the morning. By the time we had them transferred, tagged, cataloged, recorded with pictures, and built new racks to hold them, it was well past seven o'clock in the evening. Most of us didn't even stop to eat. We were left shaking and dehydrated after seven hours in a holding tank.

The next shipment was similar, and then we got a message to hold off. This allowed us enough time to try to sleep, but not enough to get actual rest. We were delusional and losing hope rapidly, which made me crack even worse dad jokes and puns. I don't know if it killed morale or helped it. Between the heat and the shipments, these corals were barely clinging on. Their polyps fully retracted, and a steady ooze of mucus covered their tissue. Their health had to be closely monitored. I was normally the first one in the tanks, making notes, taking photos of any corals that may be eventually lost, and reporting inventory of what we had. Like I mentioned earlier, coral have a threshold tolerance for heat before kicking out their zooxanthellae, another marker of stress that happens with coral, and that causes the extra production of mucus.

By the fifth day, we had received all of the coral shipments and were completely wrung out. Just because we were there unloading everything didn't mean that the online meetings stopped. A wide variety of news outlets covered the bleaching event and were contacting us for commentary and interviews. We ignored them. We didn't have time to stop working and talk about our living nightmare. We worked. Then we worked more. Slept. Ate. Worked. And worked some more. There is no such thing as a work-life balance when you're in the middle of an environmental crisis.

The sixth day, we finally established a sense of normalcy. We implemented a more organized system with roles specifying who was doing what, yet the chaos of the day was to be expected. A lot of it consisted of mustering some positive thinking and relying on each other amidst the delirium. We operated with some blind hope that all of this effort wasn't for nothing in the end. It wasn't.

"Less than 22% of the about 1,500 outplanted staghorn coral remained alive across five of the seven Mission: Iconic Reef sites surveyed...

> A subsequent update showed that less than 5% of the about 1,000 outplanted elkhorn coral the experts surveyed were alive... Without intervention by coral practitioners this past summer, the region may have completely lost the last remaining wild elkhorn and staghorn corals in the Florida Keys ecosystem."
>
> (Thiem, 2024)

So, no. What we did in relocating all of those corals wasn't the best answer, and what we did in outplanting wasn't the best answer either.

> "The deaths amounted to a 77% loss in genetic diversity needed to help sustain a vanishing coral that once blanketed Florida reefs."
>
> (Staletovich, 2024)

This event and others like it have pushed researchers on the brink themselves, arguing that restoration and outplanting may no longer be a viable solution. I argue this point of view. Coral reefs take thousands of years to develop. Most researchers are already hell-bent on learning from past experiences and constantly developing new tactics to save these endangered species. Some groups are already researching and genotyping the specific species that were unaffected by the bleaching event, and along some areas untouched at all. The most pressing question is *why*?

With others creating gene banks to create a Noah's Ark of coral genotypes, it is a natural human desire to expect to see a result of the hard work done today, tomorrow. That's just not how coral restoration works. Even if we completely shut off all gas emissions tomorrow, we would still see effects in the coming years. We need to be patient to see the results of the work we've done to date. We may not see it now, but we will see it.

Chapter 14

Here's to Hope

I t is devastating to see what inspired you, what taught you, and what is so integral to who you are as a human vanish before your eyes.

Corals have existed on our planet for thousands of years and yet we are losing them at a warp speed pace. We are seeing the reports in our news, in our media, and in our backyards. It has left many of us scientists feeling lost and even more full of questions. *Is it even worth it anymore? What can we do about this? Is this the end?* To that I say: *Hold on.* We need to hold on to our determination, hold on to our hope, and hold on to our passion.

As scientists, we knew we were walking into a hard battle. We knew and we did it anyway because we believe in our work. We believe that doing something is better than doing nothing. We believe that what we do now will help in the future. In the middle of a crisis, we often find the most inspiration. We see the organizations, brilliant minds, and people across the world coming together. We are still here, and there are thousands of us. Divers. Volunteers. Parents. Kids. Researchers. Students. Engineers. Interns. We are all working tirelessly, around the clock, to do what we can to make a difference. Scientists before us have undoubtedly had losses, and yet they persevered for the hope of the future. Now it's our turn. But we cannot do it without you.

> "Approximately 500 million people worldwide depend upon reefs for food and their livelihoods, and thirty million are almost totally dependent upon reefs. The coral reefs are supporting people whose lives depend on these natural resources for a source of food and income. Over 39% of the world population now live within 100 kilometers of the coast and many people in these areas depend on reefs."
>
> (Van den Hoek & Bayoumi, 2017)

Coral researchers do not depend on coral reefs just for our lifeblood. We don't do this job for the paycheck. We don't do it solely out of pure compassion and empathy for an ecosystem and our planet. We do it because we see something incredibly larger than ourselves. Corals are so much more. Coral reefs are so much more. The ocean is so much more. Can you put a price on seeing an octopus for the first time? Can you put a price on driving to the beach and seeing dolphins while you watch the sunset? Can you put a price on scuba diving underwater? Can you put a price on seeing your kid achieve their dreams? On their laugh while on a boat? Gifting a friend a shell you found on the beach?

Doesn't the ecosystem that provides so much deserve just a little of your attention? Sure, if you really wanted to prove a point, you could calculate the cost of contributing financially. The gas spent, the time spent, the ticket to get into the aquarium, or the parking pass at the beach. But at the end of the day, you can't price out the time. The memories. The overwhelming human experience.

The next time you look at the ocean, I want you to think of coral, the little underrated polyp creature that is the backbone of our oceans. I want you to think of us, the people that work so hard that their life is their work. People that would much rather be underwater than above it. People that get just as tired, frustrated, happy, sleepy, and hungry as you. The thousands upon thousands of people that deeply care and are fighting every day to protect a foundational piece of ourselves. Each thinking, *If not me, then who?*

I want you to care. To care about our researchers and scientists, enough to listen to us. To care enough about our oceans to share what you've learned. To care about curing coral.

Chapter 15

Caring About Coral

The vibrant, living tapestry of coral reefs tells a story of resilience and interconnectedness. Yet, it is a story at risk of being silenced by the cumulative weight of human activity. These ancient underwater civilizations, which support 25% of marine species and protect coastal communities, are facing threats from climate change, pollution, overfishing, and more. Saving coral reefs isn't just an environmental imperative; it is a necessity for the health of our planet and our own survival.

This chapter is a call to action, outlining both grand and humble ways to protect our reefs and preserve the environment, empowering individuals, communities, and organizations to make a tangible impact. Besides the few suggestions at the end of the previous chapters for some small actions you can take to help our oceans, there is one universal action you can take that will help no matter what your specific focus is when it comes to our planet. That is to first and foremost believe us. We don't do this job for the recognition, the praise, or any egotistical basis. We do it because we care and we believe in a better future for our world. So, believe the scientists yelling from the rooftops, telling you—begging you—to consider that something is wrong.

Then, add your voice. There are many ways to contribute to the conversation around our planet and our oceans. Advocacy is one of the most powerful tools. Please write to your county-level, state-level, and even national representatives. Writing to lawmakers and attending town halls to support policies that address climate change, regulate harmful practices, and reduce marine pollution can drive systemic change. If you have no interest in policy, educators and influencers can amplify these efforts by teaching others about coral reefs and the threats they face. Whether in classrooms, community centers, or online platforms, spreading awareness of what is happening in the world around us fosters a culture of conservation.

Saving coral reefs and safeguarding our planet requires a collective effort from all of us. While the challenges are immense, with the help of scientists and you, the solutions are beginning to be within reach. Each of us has a role to play—whether by advocating for large-scale changes, taking small everyday actions, being a hands-on scientist, volunteering in local conservation projects, involving yourself in environmental policy, or supporting organizations that lead the charge. When it comes to tackling the crisis facing coral reefs, global-scale actions are critical. Reducing greenhouse gas emissions is perhaps the most urgent need. Climate change causes ocean warming and acidification, which severely harm coral reefs.

Transitioning to various alternate energy resources is essential. Whether it's advocating for government policies that prioritize solar, wind, and hydroelectric power, or supporting businesses that invest in clean energy technologies, everyone can help accelerate the shift away from fossil fuels. Another large-scale action is establishing and enforcing marine protected areas (MPAs). MPAs serve as safe havens for marine life, allowing ecosystems to recover and thrive. Supporting international initiatives like the "30 x 30" goal, which aims to protect 30% of the ocean by 2030, can make a transformative difference. Similarly, ending harmful industrial practices, such as deep-sea mining and unregulated fishing, can prevent widespread destruction of marine habitats. Advocating for global treaties that combat plastic pollution, like the United

Nations' Global Plastics Agreement, also has far-reaching implications for coral reef conservation.

While these global initiatives are vital to see widespread change, small individual actions play an equally important role in the fight to save coral reefs. There are hundreds of thousands of different environmental conservation organizations globally, working toward a collective goal: to make the planet a better place. You can volunteer your time toward their mission. They do not have too little work on their hands and can use all the help that they can get. It does not necessarily need to be a marine-focused program, either; any step toward the stewardship and protection of our planet is the right one.

Find a local environmental program or a program that you are passionate about—from orcas to salmon, octopuses to corals—and donate there, whether it be your time or any spare funding. If you are limited on personal time, you can begin your journey in your home. Start by reducing your carbon footprint: switching to energy-efficient appliances, using LED light bulbs, and being mindful of energy consumption at home can make a significant impact over time. Another meaningful step is minimizing waste. Say no to single-use plastics like straws, bags, and bottles, and instead opt for reusable alternatives. You can also consider your waste in your home and try composting.

Composting reduces some of the drain of resources, transforming your food scraps into enriched soil that can then be donated or used in your very own garden.

> "In the United States, food waste is estimated at between 30–40% of the food supply. This is based on USDA estimates of 31% food loss at the retail and consumer levels. This added up to approximately 133 billion pounds and $161 billion worth of food in 2010."
>
> (U.S. Department of Agriculture, n.d.)

Supporting the environment can even transfer to your plate. You can make sustainable seafood choices by opting for fish that are harvested responsibly or farmed sustainably, which can also alleviate pressure on marine ecosystems. Additionally, supporting businesses that prioritize eco-friendly practices reinforces market demand for sustainable options. The key is to take any action, knowing that every step forward contributes to a larger wave of change.

When it comes to protecting coral reefs directly, using reef-safe sunscreen is a simple yet effective measure. Traditional sunscreens often contain harmful chemicals like oxybenzone and octinoxate, which can bleach and damage corals. The next time you are on vacation on a snorkel trip, mention how you once read a book that said *please, please, please do not touch the corals*. They are animals. They are sensitive, and they won't like it. Any time you are at the beach, bring a small bag with you to clean up the trash you find on the shoreline.

Without coral, we won't have coral reefs. If you have learned anything in this book, remember that coral reefs are more than just underwater marvels; they are integral to the planet's life-support system. Their survival is tied to ours, and the time to act is now. Together, we can turn the tide and ensure that coral reefs—and the planet as a whole—continue to flourish for generations to come. Our actions today will shape the legacy we leave behind. Let's make it a legacy worth remembering and let the generations after us know that we cared.

P.S.

I am so thankful that my book ended up in your hands. I truly hope that you found it entertaining, informative, and relatable. We are all just humans trying to make the world a better place in our own way.

This story is the result of connections made with people from all around the world, and that can include you as well. Along these lines, I personally invite you to come check out the growing ocean community online via Instagram under the handle @summerscientist. There, I continue my ocean education focus and share tons of photos behind the scenes of conducting research and outreach, wherever it takes me.

Like a good scientist, I also welcome all responses, reactions, and data. If you enjoyed the adventure or have any questions, please post a review of this book on Amazon! And feel free to reach out to me, either directly via email (thesummerscientist@gmail.com) or through my website (summerscientist.com), where I share podcasts I've been on, additional marine science student help, and tons of other helpful content. If you didn't like it, well, at least you didn't live it, and you can donate this book out of your house and your head.

References

Almroth, B. C., & Eggert, H. (2019). Marine plastic pollution: Sources, impacts, and policy issues. *Review of Environmental Economics and Policy*, *13*(2), 317–326. https://doi.org/10.1093/reep/rez012

Cardenas, J. A., Samadikhoshkho, Z., Rehman, A. U., Valle-Pérez, A. U., Herrera-Ponce de León, E., Hauser, C. A. E., Feron, E. M., & Ahmad, R. (2024). A systematic review of robotic efficacy in coral reef monitoring techniques. *Marine Pollution Bulletin*, *202*, 116273. https://doi.org/10.1016/j.marpolbul.2024.116273

Casoli, E., Ventura, D., Mancini, G., Pace, D. S., Belluscio, A., & Ardizzone, G. (2021). High spatial resolution photo mosaicking for the monitoring of coralligenous reefs. *Coral Reefs*, *40*, 1267–1280. https://doi.org/10.1007/s00338-021-02136-4

Doney, S. C., Fabry, V. J., Feely, R. A., & Kleypas, J. A. (2009). Ocean acidification: The other CO2 problem. *Annual Review of Marine Science*,

1, 169–192. https://doi.org/10.1146/annurev. marine.010908.163834

Donovan, C., Towle, E. K., Kelsey, H., Allen, M., Barkley, H., Besemer, N., Blondeau, J., Eakin, M., Edwards, K., Enochs, I. C., Fleming, C., Geiger, E., Grove, L. J., Groves, S., Johnson, M., Johnston, M., Kindinger, T., Manzello, D., Miller, N., … Viehman, S. (2020). *Coral reef condition: A status report for U.S. coral reefs.* NOAA Coral Reef Conservation Program and University of Maryland Center for Environmental Science. https://doi.org/10.25923/wbbj-t585

Eaton, K. R., Landsberg, J. H., Kiryu, Y., Peters, E. C., & Muller, E. M. (2021). Measuring stony coral tissue loss disease induction and lesion progression within two intermediately susceptible species, *Montastraea cavernosa* and *Orbicella faveolata*. *Frontiers in Marine Science, 8,* 717265. https://doi. org/10.3389/fmars.2021.717265

Ferrari, R., Lachs, L., Pygas, D. R., Humanes, A., Sommer, B., Figueira, W. F., Edwards, A. J., Bythell, J. C., & Guest, J. R.(2021). Photogrammetry as a tool to improve ecosystem restoration. *Trends in Ecology & Evolution, 36*(12), 1093–1101. https:// doi.org/10.1016/j.tree.2021.07.004

Forsman, Z. H., Page, C. A., Toonen, R. J., & Vaughan, D. (2015). Growing coral larger and faster: Micro-colony-fusion as a strategy for accelerating coral cover. *PeerJ Life & Environment, 3,* e1313. https://doi.org/10.7717/peerj.1313

Goreau, T. J. F. (2022). Perspective chapter: Electric reefs enhance coral climate change adaptation. In G. Chimienti (Ed.), *Corals: Habitat formers in the Anthropocene.* IntechOpen. https://doi.org/10.5772/intechopen.107273

Guo, W., Bokade, R., Cohen, A. L., Mollica, N. R., Leung, M., & Brainard, R. E. (2020). Ocean acidification has impacted coral growth on the great barrier reef. *Geophysical Research Letters, 47*(19), e2019GL086761. https://doi.org/10.1029/2019GL086761

Hartmann, N. B., Hüffer, T., Thompson, R. C., Hassellöv, M., Verschoor, A., Daugaard, A. E., Rist, S., Karlsson, T., Brennholt, N., Cole, M., Herrling, M. P., Hess, M. C., Ivleva, N. P., Lusher, A. L., & Wagner, M. (2019). Are we speaking the same language? Recommendations for a definition and categorization framework for plastic debris. *Environmental Science & Technology, 53*(3), 1039–1047. https://pubs.acs.org/doi/10.1021/acs.est.8b05297

Keller, K. A., Innis, C. J., Tlusty, M. F., Kennedy, A. E., Bean, S. B., Cavin, J. M., & Merigo, C. (2012).

Metabolic and respiratory derangements associated with death in cold-stunned Kemp's ridley turtles (*Lepidochelys kempii*): 32 cases (2005–2009). *Journal of the American Veterinary Medical Association, 240*(3), 317–323. https://doi.org/10.2460/javma.240.3.317

Klepac, C. N., Petrik, C. G., Karabelas, E., Owens, J., Hall, E. R., & Muller, E. M. (2024). Assessing acute thermal assays as a rapid screening tool for coral restoration. *Scientific Reports, 14*, 1898. https://doi.org/10.1038/s41598-024-51944-5

Koch, H. R., Matthews, B., Leto, C., Engelsma, C., & Bartels, E. (2022). Assisted sexual reproduction of *Acropora cervicornis* for active restoration on Florida's Coral Reef. *Frontiers in Marine Science, 9*, 959520. https://doi.org/10.3389/fmars.2022.959520

Leonard, S. V. L., Liddle, C. R., Atherall, C. A., Chapman, E., Watkins, M., Calaminus, S. D. J., & Rotchell, J. M. (2024). Microplastics in human blood: Polymer types, concentrations and characterisation using µFTIR. *Environment International, 188*, 108751. https://doi.org/10.1016/j.envint.2024.108751

Lindsey, R., & Dahlman, L. (2025). *Climate change: Global temperature*. National Oceanic and Atmospheric Administration. https://www.

climate.gov/news-features/understanding-climate/climate-change-global-temperature

Miller, M., Bourque, A., & Bohnsack, J. (2002). An analysis of the loss of Acroporid corals at Looe Key, Florida, USA: 1983–2000. *Coral Reefs, 21*(2), 179–182. https://doi.org/10.1007/s00338-002-0228-7

Miller, M. E., Motti, C. A., Hamann, M., & Kroon, F. J. (2023). Assessment of microplastic bioconcentration, bioaccumulation and biomagnification in a simple coral reef food web. *Science of the Total Environment, 858*(1), 159615. https://doi.org/10.1016/j.scitotenv.2022.159615

Mou, S., Tsai, D., & Dunbabin, M. (2022). Reconfigurable robots for scaling reef restoration. arXiv preprint arXiv:2205.04612

National Oceanic and Atmospheric Administration. (n.d.). *Cold-stunning and sea turtles – Frequently asked questions.* NOAA Fisheries. https://www.fisheries.noaa.gov/national/marine-life-distress/cold-stunning-and-sea-turtles-frequently-asked-questions

National Oceanic and Atmospheric Administration. (n.d.). *What is coral bleaching?* National Ocean Service. https://oceanservice.noaa.gov/facts/coral_bleach.html

Neely, K. L., Macaulay, K. A., Hower, E. K., & Dobler, M. A. (2020). Effectiveness of topical antibiotics in treating corals affected by stony coral tissue loss disease. *PeerJ Life & Environment, 8*, e9289. https://doi.org/10.7717/peerj.9289

Neely, K. L., Nowicki, R. J., Dobler, M. A., Chaparro, A. A., Miller, S. M., & Toth, K. A. (2024). Too hot to handle? The impact of the 2023 marine heatwave on Florida Keys coral. *Frontiers in Marine Science, 11*, 1489273. https://doi.org/10.3389/fmars.2024.1489273

Niemuth, J. N., Harms, C. A., Macdonald, J. M., & Stoskopf, M. K. (2020). NMR-based metabolomic profile of cold stun syndrome in loggerhead (*Caretta caretta*), green (*Chelonia mydas*), and Kemp's ridley (*Lepidochelys kempii*) sea turtles in North Carolina, USA. *Wildlife Biology, 2020*(1). https://doi.org/10.2981/wlb.00587

O'Neill, S. (2022). Solar geoengineering to reduce global warming—The outlook remains cloudy. *Engineering, 9*, 6–9. https://doi.org/10.1016/j.eng.2021.12.005

Page, C. A., Muller, E. M., & Vaughan, D. E. (2018). Microfragmenting for the successful restoration of slow growing massive corals. *Ecological*

Engineering, 123, 86–94. https://doi.org/10.1016/j.ecoleng.2018.08.017

Page, C., & Vaughan, D. E. (2014, March). *The cultivation of massive corals using "micro-fragmentation" for the "reskinning" of degraded coral reefs* [Poster presentation]. Benthic Ecology Meeting, University of North Florida, Jacksonville, FL, United States. https://www.researchgate.net/publication/287204400_The_cultivation_of_massive_corals_using_micro-fragmentation_for_the_reskinning_of_degraded_coral_reefs

Piñeros, V. J., Reveles-Espinoza, A. M., & Monroy, J. A. (2024). From remote sensing to artificial intelligence in coral reef monitoring. *Machines, 12*(10), 693. https://doi.org/10.3390/machines12100693

Price, J. T., Drye, B., Domangue, R. J., & Paladino, F. V. (2018). Exploring the role of artificial lighting in loggerhead turtle (*Caretta caretta*) nest-site selection and hatchling disorientation. *Herpetological Conservation and Biology, 13*(2), 415–422.

Richmond, R. H., Tisthammer, K. H., & Spies, N. P. (2018). The effects of anthropogenic stressors on reproduction and recruitment of corals and reef

organisms. *Frontiers in Marine Science, 5,* 226.
https://doi.org/10.3389/fmars.2018.00226

Schuyler, Q. A., Wilcox, C., Townsend, K. A.,
Wedemeyer-Strombel, K. R., Balazs, G., van
Sebille, E., & Hardesty, B. D. (2015). Risk analysis
reveals global hotspots for marine debris ingestion
by sea turtles. *Global Change Biology, 22*(2), 567–
576. https://doi.org/10.1111/gcb.13078

Staletovich, J. (2024, August 20). A "catastrophe" in
the Lower Keys: Summer heatwave wipes out
iconic elkhorn coral. WUSF Public Media. https://
www.wusf.org/environment/2024-08-20/summer-
heatwave-wipes-out-elkhorn-coral-lower-keys

Thiem, H. (2024). *The future of coral restoration in
the Florida Keys after unprecedented marine heat
wave of 2023.* National Oceanic and Atmospheric
Administration. https://prod-01-asg-www-climate.
woc.noaa.gov/news-features/event-tracker/
future-coral-restoration-florida-keys-after-
unprecedented-marine-heat

Ullah, S., Ahmad, S., Guo, X., Ullah, S., Nabi, G.,
& Wang, K. (2023). A review of the endocrine
disrupting effects of micro and nano plastic and
their associated chemicals in mammals. *Frontiers
in Endocrinology, 13,* Article 1084236. https://doi.
org/10.3389/fendo.2022.1084236

U.S. Department of Agriculture. (n.d.). *Why should we care about food waste?* USDA. https://www.usda. gov/about-food/food-safety/food-loss-and-waste/ why-should-we-care-about-food-waste

Van den Hoek, L. S., & Bayoumi, E. K. (2017). Importance, destruction and recovery of coral reefs. *IOSR Journal of Pharmacy and Biological Sciences, 12*(2), 59–63. https://doi. org/10.9790/3008-1202025963

Voolstra, C. R., Peixoto, R. S., & Ferrier-Pagès, C. (2023). Mitigating the ecological collapse of coral reef ecosystems: Effective strategies to preserve coral reef ecosystems. *EMBO Reports, 24,* e56826. https://doi.org/10.15252/embr.202356826

Webster, F. J., Babcock, R. C., Van Keulen, M., & Loneragan, N. R. (2015). Macroalgae inhibits larval settlement and increases recruit mortality at Ningaloo Reef, Western Australia. *PLOS One, 10*(4), e0124162. https://doi.org/10.1371/journal. pone.0124162

Acknowledgments

I consider myself an extremely lucky and fortunate person in this life. I have had the overwhelming privilege to work alongside, and make lifelong friends with, some of the world's best scientists in their field. I am continually impressed, inspired, and simply awestruck by their dedication and who they are as human beings. This book is nothing but a mere glimpse into the amount of knowledge, research, effort, and passion each one of them holds. I am honored to have learned from and worked with such individuals. To my interns, who fostered my love for teaching and sharing in a way that I never thought would be possible: you taught me so much more than I could ever teach you, and for that I am eternally grateful.

My thanks also go to those who have cheered me on and helped foster the human I am today. To my dear friend Chloe Spring, who listened to me read chapters back to her over and over to the point that it made us both dizzy. To the love of my life, Dylan, who has cooked many dinners when I was too wrapped up in writing to notice what time it was, fed me ice cream when I needed a break, and has always, always loved me tenfold. I love you more. I win. It's in a published book all over the world now.

To my family. My parents, grandparents, and brother: you have each always cheered me on from when I could not even pronounce "marine biologist" correctly, allowed me to watch *Shark Week* reruns way too much, and taught me not to take shit from anyone, including myself. Here's to making good decisions.

To my sweet fur boys Milo and Chowder, who snuggled me the whole way through. To Bonnie and Jack, you helped me when I was the most clueless. You encouraged me and helped me become a young professional, and I would not have gotten to where I am without your guidance. Allison, you taught me to shoot for the stars and never look back. Thank you for encouraging me to start that silly little Instagram page.

To my editor, Hugo Villabona, and the entire Mango Publishing team, who kindly guided me through the writing process and have been nothing but a dream to work with through each step, including editing, design, and publishing this book. This would not have been possible *at all* without you. You undoubtedly read my first draft and were terrified at all of the grammatical mistakes. :)

About the Author

As a scientist, educator, and now author, Summer Collins has dedicated her life to the ocean. Funnily enough, she never intended on becoming a teacher, and amongst her many hats of being a coral scientist, scuba diver, online science communicator, and director of education, being a teacher is the one she enjoys wearing the most. Collins received her degree from University of North Carolina Wilmington in environmental science with a minor in biology and a focus in conservation. She began her career at Mote Marine Laboratory after several years of various internships and other research projects.

During her time working as a coral biologist, Summer began a science communication page to further share her passion and love of corals while researching endangered Florida coral species. That page has grown immensely, allowing her to also provide scientific trips and one-on-one coaching opportunities for students.

She has since gone on to work for several different agencies, broadening her scope in the realm of marine science from working in the coral aquarium aquaculture industry to now being an educator for a nonprofit. Summer Collins's lighthearted, "anyone can be a marine biologist" approach to science allows many to learn from her and has led to numerous opportunities, such as being featured in a kid's book, on various podcasts, and having her lab's work appear on CNN, BBC, and other worldwide news outlets.

That being said, Summer is most proud when hearing from various students around the world and encouraging the success of her former interns. Summer aims to obtain her master's degree, continuing her focus in marine science. Her lifelong goal is to one day start her own nonprofit, with the aim of properly supporting students while they gain hands-on experience in the field. With an adventurous spirit, Summer currently travels the country for various biology-based roles with her husband, Dylan; rescue pitbull, Milo; and cat, Chowder.